普通高等教育新工科人才培养规划教材（大数据专业）

Hadoop 大数据开发

主　编　刘春阳　张学龙　刘丽军

副主编　陈　勇　陈艳男　蒋中洲　王宇希

中国水利水电出版社
www.waterpub.com.cn
·北京·

内容提要

本书通过原理加案例方式系统讲解了 Hadoop 大数据开发，精心安排了原理分析、环境搭建、案例开发等环节，使读者对解决大数据问题有清晰的思路。

全书共 7 章：前 6 章系统讲解大数据 Hadoop 架构，包括大数据处理平台 Hadoop、分布式文件系统 HDFS、并行计算模型 MapReduce、资源调度框架 Yarn；第 7 章是 MapReduce 应用实例，通过案例帮助读者进一步理解 Hadoop 平台。全书突出三个特点：道理简单明了、思路清晰透彻、案例新颖实用。

本书可作为普通高校大数据相关专业的教材，可供想深入了解 Hadoop 架构编程的读者参考，还可作为相关培训班的培训教材。

图书在版编目（CIP）数据

Hadoop大数据开发 / 刘春阳，张学龙，刘丽军主编
. -- 北京：中国水利水电出版社，2018.9
普通高等教育新工科人才培养规划教材. 大数据专业
ISBN 978-7-5170-6903-4

Ⅰ. ①H… Ⅱ. ①刘… ②张… ③刘… Ⅲ. ①数据处理软件－高等学校－教材 Ⅳ. ①TP274

中国版本图书馆CIP数据核字(2018)第216857号

策划编辑：石永峰　　　责任编辑：张玉玲　　　封面设计：梁 燕

书　名	普通高等教育新工科人才培养规划教材（大数据专业） **Hadoop 大数据开发** Hadoop DASHUJU KAIFA
作　者	主　编　刘春阳　张学龙　刘丽军 副主编　陈　勇　陈艳男　蒋中洲　王宇希
出版发行	中国水利水电出版社 （北京市海淀区玉渊潭南路1号D座　100038） 网址：www.waterpub.com.cn E-mail: mchannel@263.net（万水） 　　　　sales@waterpub.com.cn 电话：（010）68367658（营销中心）、82562819（万水）
经　售	全国各地新华书店和相关出版物销售网点
排　版	北京万水电子信息有限公司
印　刷	三河市鑫金马印装有限公司
规　格	184mm×260mm　16开本　11.5印张　280千字
版　次	2018年9月第1版　2018年9月第1次印刷
印　数	0001—4000 册
定　价	32.00 元

凡购买我社图书，如有缺页、倒页、脱页的，本社营销中心负责调换
版权所有·侵权必究

前　　言

这是一个大数据爆发的时代，面对信息的激流、多元化数据的涌现，大数据已经为个人生活、企业经营，甚至国家与社会的发展带来了机遇和挑战，成为信息产业中极具潜力的增长点。大数据时代在众多领域掀起变革的巨浪，但我们要冷静地看到，大数据的核心在于为客户挖掘数据中蕴藏的价值，而不是软硬件简单地堆砌。因此，针对不同领域的大数据应用模式、商业模式研究将是大数据产业健康发展的关键。

Hadoop 技术能够成功的最根本原因在于它是把传统的集中式运算转化成分布式计算的一种有效手段。Hadoop 的分布式文件系统能够以可靠快捷的方式将数据分布存储到不同计算节点中，Hadoop MapReduce 编程又能够以简单的方法为人们提供分布式编程接口，从而降低了分布式开发门槛。

本书共 7 章，不仅有详细的理论讲解，还有大量的实战操作，具体内容如下：

第 1 章深入探究大数据的概念、产生的背景和发展现状，应用案例指出了大数据面临的机遇与挑战，介绍大数据处理技术和计算模式，最后阐述大数据与云计算之间的区别和联系。

第 2 章详细介绍大数据处理平台 Hadoop 的生态系统和架构。

第 3 章讲解 Hadoop 分布式平台的搭建和验证。

第 4 章描述 HDFS 的架构、工作机制、文件读写流程和 Shell 命令。

第 5 章讲解 HDFS Windows 远程开发、HDFS Java API 接口和编程实战。

第 6 章讲解 MapReduce 编程模型、工作原理和 Yarn 资源管理。

第 7 章讲解常用的 MapReduce Java API 接口、应用实例和高级编程。

本书的编写得到北京百知教育科技有限公司的大力支持，在此表示感谢。

由于时间仓促及编者水平有限，本书难免存在不足之处，恳请读者批评指正。

编　者

2018 年 7 月

目 录

前言

第1章 大数据概论 ... 1
1.1 大数据概述 ... 1
1.1.1 大数据产生的时代背景 ... 1
1.1.2 大数据的特征 ... 2
1.1.3 大数据应用案例 ... 2
1.1.4 大数据的机遇与挑战 ... 5
1.2 大数据处理技术 ... 5
1.3 大数据与云计算 ... 6
1.4 本章小结 ... 7

第2章 大数据处理平台Hadoop ... 8
2.1 Hadoop生态系统 ... 8
2.2 Hadoop架构 ... 11
2.2.1 HDFS ... 12
2.2.2 MapReduce ... 12
2.2.3 Yarn ... 13
2.3 Hadoop版本变迁 ... 13
2.3.1 Hadoop发展史 ... 13
2.3.2 如何选择Hadoop开发版本 ... 14
2.4 本章小结 ... 14

第3章 Hadoop平台搭建 ... 15
3.1 基础环境配置 ... 15
3.2 Hadoop配置文件修改 ... 15
3.3 Hadoop平台运行及验证 ... 22
3.4 本章小结 ... 23

第4章 分布式文件系统HDFS ... 24
4.1 HDFS架构 ... 24
4.1.1 HDFS的基本框架 ... 24
4.1.2 HDFS的特点 ... 26
4.2 HDFS的工作机制 ... 27
4.2.1 HDFS读写过程分析 ... 27
4.2.2 NameNode的工作机制 ... 29
4.2.3 元数据的CheckPoint ... 32
4.2.4 DataNode的工作机制 ... 33
4.3 HDFS shell命令 ... 34
4.3.1 帮助相关命令 ... 35
4.3.2 查看相关命令 ... 36
4.3.3 文件及目录相关命令 ... 37
4.3.4 统计相关命令 ... 46
4.3.5 快照命令 ... 47
4.4 本章小结 ... 48

第5章 HDFS Java API编程 ... 49
5.1 远程开发环境搭建 ... 49
5.2 HDFS Java API接口 ... 53
5.3 HDFS Java API编程 ... 53
5.3.1 获取文件系统 ... 55
5.3.2 列出所有DataNode的名字信息 ... 56
5.3.3 创建文件目录 ... 57
5.3.4 删除文件或文件目录 ... 58
5.3.5 查看文件是否存在 ... 59
5.3.6 文件上传至HDFS ... 59
5.3.7 从HDFS下载文件 ... 60
5.3.8 文件重命名 ... 61
5.3.9 遍历目录和文件 ... 62
5.3.10 根据filter获取目录下的文件 ... 63
5.3.11 取得数据块所在的位置 ... 65
5.4 程序打包 ... 66
5.5 本章小结 ... 68

第6章 并行计算MapReduce ... 69
6.1 MapReduce编程模型 ... 69
6.1.1 并行编程模型概述 ... 69
6.1.2 并行计算编程模型 ... 70
6.1.3 MapReduce编程模型 ... 72
6.2 MapReduce工作原理 ... 73
6.3 Yarn ... 75
6.3.1 Yarn基本框架与组件 ... 75
6.3.2 Yarn工作流程 ... 76

6.3.3 新旧 Hadoop MapReduce 框架对比 …… 77
6.4 MapReduce Shuffle 性能调优 …………… 79
6.5 本章小结 ……………………………………… 80
第 7 章 MapReduce Java API 编程 …………… 81
 7.1 MapReduce Java API 接口讲解 ………… 81
  7.1.1 InputFormat 接口 ……………………… 82
  7.1.2 Mapper 类 ……………………………… 85
  7.1.3 Partitioner 类 …………………………… 87
  7.1.4 Combiner 类 …………………………… 88
  7.1.5 Reducer 类 ……………………………… 89
  7.1.6 OutputFormat 接口 …………………… 90
  7.1.7 GenericOptionsParser 类 …………… 91
  7.1.8 DistributedCache 类 ………………… 91
 7.2 MapReduce Java API 应用实例 ………… 92
  7.2.1 统计单词出现频率 ……………………… 92
  7.2.2 统计出现的单词 ………………………… 96
  7.2.3 统计平均成绩 …………………………… 99
  7.2.4 排序 ……………………………………… 101
  7.2.5 求年最高温度 …………………………… 103
  7.2.6 关系运算——投影运算 ………………… 106
  7.2.7 关系运算——并运算 …………………… 108
  7.2.8 关系运算——交运算 …………………… 110
  7.2.9 关系运算——差运算 …………………… 111
  7.2.10 关系运算——连接运算 ……………… 114
 7.3 MapReduce Java API 高级编程 ………… 116
  7.3.1 多输入路径方式 ………………………… 116
  7.3.2 使用 Partitioner 实现输出到多个文件 …………………………………………… 119
  7.3.3 自定义 OutputFormat 文件输出 …… 122
  7.3.4 文本文件转化成 XML 文件 …………… 127
  7.3.5 通过 MultipleOutputs 完成多文件输出 ……………………………………… 130
  7.3.6 将 MapReduce 产生的结果集导入到 MySQL 中 …………………………… 135
  7.3.7 自定义比较器 …………………………… 140
  7.3.8 MapReduce 分析明星微博数据 ……… 145
  7.3.9 MapReduce 最佳成绩统计 …………… 152
  7.3.10 MapReduce 链接作业 ……………… 158
  7.3.11 利用 Job 嵌套求解二度人脉 ………… 162
 7.4 本章小结 ……………………………………… 168
附录 CentOS7 安装 ……………………………… 169

第 1 章　大数据概论

随着互联网的飞速发展，数据已经积累到了一个由量变引起质变的程度，大数据（Big Data）几乎应用到了人们发展的所有领域中，不管是云计算、物联网，还是社交网络、移动互联网等都会与大数据扯上关系。那么，什么是大数据，大数据发展的现状如何，大数据能给人们带来什么？通过本章的学习，您将会得到答案。

1.1　大数据概述

网络和信息技术的不断发展，带动了移动设备和通信手段（如社交网站）的革新，人类生产的数据量每年都在快速增长。各行业信息化程度的提高导致业务数据正以几何级数的形式爆发，预计到2020年全球将总共拥有35亿GB的数据量，其收集、存储、格式、检索、分析、应用等存在诸多问题，不能再以传统的信息处理技术加以解决。

目前对大数据的准确定义尚有一些争论，这就导致大数据的定义有多种。维基百科给出的定义是：大数据是利用常用软件工具捕获、管理和处理数据所耗时间超过可容忍时间的数据集。美国国家科学基金会（NSF）则将大数据定义为"由科学仪器、传感设备、互联网交易、电子邮件、音视频软件、网络点击流等多种数据源生成的大规模、多元化、复杂、长期的分布式数据集"。全球知名的咨询公司麦肯锡认为：大数据是指无法在一定时间内用传统数据库软件工具对其内容进行采集、存储、管理和分析的数据集合。但它同时指出"大数据"并非总是说有数百个TB才得上，根据实际使用情况，有时候数百个GB的数据也可以称为大数据，这主要看它的第三个维度，也就是速度或者时间维度。

我国政府、产业界和学术界也做了相应的理论研究和实践研究。2015年9月，国务院印发《促进大数据发展行动纲要》，系统部署大数据发展工作。2016年3月17日，《中华人民共和国国民经济和社会发展第十三个五年规划纲要》发布，其中第二十七章"实施国家大数据战略"提出：把大数据作为基础性战略资源，全面实施促进大数据发展行动，加快推动数据资源共享开放和开发应用，助力产业转型升级和社会治理创新。具体包括：加快政府数据开放共享、促进大数据产业健康发展。

1.1.1　大数据产生的时代背景

随着计算机存储能力的提升和复杂算法的发展，近年来的数据量成指数型增长，这些趋势也使科学技术发展日新月异，商业模式发生了颠覆式变化。数据正在被商业化，来自网络、智能手机、传感器、嵌入式设备以及其他途径的数据形成了一项资产，产生了巨大的商业价值。苹果、亚马逊、Facebook、谷歌、阿里巴巴等利用大数据分析自己的优势，改变了竞争的基础，建立了全新的商业模式。稀缺数据的所有者利用数字化网络平台在一些市场近乎垄断，只需用独特的方式将数据整合分析，提供有价值的数据分析。2011年全球的数据存储量就达到1.8ZB，与2011年相比2015年大数据增长了近4倍，未来十年，全球数据存储量还将增长十倍,大数据成为提升产业竞争力和创新商业模式的新途径。

大数据从产生到目前风靡全球,大致经历了以下3个发展阶段:

(1) 20世纪末至21世纪初:大数据的萌芽期。随着数据挖掘理论和数据库技术的逐步成熟,一批商业智能工具和知识管理技术开始被应用,如数据仓库、专家系统、知识管理系统等。

(2) 21世纪前10年:大数据的成熟期。Web2.0应用迅猛发展,非结构化数据大量产生,传统处理方法难以应对,带动了大数据技术的快速突破,大数据解决方案逐渐走向成熟,形成了并行计算与分布式系统两大核心技术,谷歌的GFD和MapReduce等大数据技术受到追捧,Hadoop平台开始大行其道。

(3) 2010年以后:大数据的大规模应用期。大数据应用渗透到各行各业,数据驱动决策,信息社会智能化程度大幅提高。

1.1.2 大数据的特征

目前,关于大数据的特征还有一定的争议,本书采用普遍被接受的4V,即规模性(Volume)、多样性(Variety)、价值密度(Value)和高速性(Velocity)进行描述。

1. 数据量大(Volume)

非结构化数据的超大规模增长导致数据集合的规模不断扩大,数据单位已经从GB级到TB级再到PB级,甚至开始以EB和ZB来计数。

2. 类型繁多(Variety)

大数据的类型不仅包括网络日志、音频、视频、图片、地理位置信息等结构化数据,还包括半结构化数据甚至是非结构化数据,具有异构性和多样性的特点。

3. 价值密度低(Value)

大数据价值密度相对较低。如随着物联网的广泛应用,信息感知无处不在,信息海量,但价值密度较低,存在大量不相关信息。因此需要对未来趋势与模式作可预测分析,利用机器学习、人工智能等进行深度复杂分析。而如何通过强大的机器算法更迅速地完成数据的价值提炼,是大数据时代亟待解决的难题。虽然单位数据的价值密度在不断降低,但是数据的整体价值在提高。

4. 速度快时效高(Velocity)

处理速度快,时效性要求高。需要实时分析而非批量式分析,数据的输入、处理和分析连贯性地处理,这是大数据区分于传统数据挖掘最显著的特征。

1.1.3 大数据应用案例

将大量的原始数据汇集在一起,通过数据挖掘等技术分析数据中潜在的规律,预测未来的发展趋势,有助于人们做出正确的决策,从而提高各个领域的运行效率,获得更大的收益。大数据冲击着许多行业,包括金融行业、互联网、医疗行业、社交网络、零售行业和电子商务等,大数据在彻底地改变着人们的生活。

1. 大数据在互联网企业的应用

互联网是最早利用大数据进行精准营销的行业,通过大数据不仅可以为企业进行精准营销,还可以快速友好地对用户实施个性化解决方案。IBM大数据提供的服务包括数据分析、文本分析、蓝色云杉（混搭供电合作的网络平台）、业务事件处理和商业化服务。基于对大数据价值的沉淀，依据信用体系等，马云将集团下的阿里金融与支付宝两项核心业务合并成立阿里小微金融；另外，为了便于在内部解决数据的交换、安全和匹配等问题，阿里集团还搭建了一个数据交换平台，在这个平台上，各个事业群可以实现数据的内部流转，实现价值最大化。

由于互联网的数据较为集中，数据量足够大，数据种类较多，因此未来互联网数据应用将会有更多的想象空间，包括预测流行趋势、消费趋势、地域消费特点、客户消费习惯、各种消费行为的相关度、消费热点、影响消费的重要因素等。

2. 大数据在医疗行业的应用

医疗行业拥有大量的病例、病理报告、治愈方案、药物报告等。如果这些数据可以被整理和应用将会极大地帮助医生和病人。人们面对的数目及种类众多的病菌、病毒，以及肿瘤细胞，都处于不断进化的过程中。在发现诊断疾病时，疾病的确诊和治疗方案的确定是最困难的。

借助于大数据平台可以收集不同病例和治疗方案，以及病人的基本特征，建立针对疾病特点的数据库。在医生诊断病人时可以参考病人的疾病特征、化验报告和检测报告，参考疾病数据库来快速帮助病人确诊，明确定位疾病。在制定治疗方案时，医生可以依据病人的基因特点，调取相似基因、年龄、人种、身体情况的有效治疗方案，制定出适合病人的治疗方案，帮助更多人及时进行治疗，同时这些数据也有利于医药行业开发出更加有效的药物和医疗器械。

3. 大数据在金融行业的应用

金融行业的数据具有交易量大、安全级别高等特点。银行在做信贷风险分析的时候，需要大量数据进行相关性分析，但是很多数据来源于政府各个职能部门，包括工商税务、质量监督、检察院法院等，这些数据短期仍然是无法拿到的。

摩根大通通过使用大数据技术以满足日益增多的需求，如诈骗检验、IT 风险管理和自助服务；存储大量非结构化数据，允许公司收集存储 Web 日志、交易数据和社交媒体数据，以方便以客户为中心的数据挖掘和数据分析工具的使用。光大银行将在线营销方案、微博营销、客户行为分析（包括电话语音、网络的监控录像等）和风险控制与管理（结构化非结构化数据整合，分析系统存在 IT 风险或者钓鱼网站防欺诈）等。建设银行充分跟进大数据时代的脚步，建立善融商务企业商城，在该平台上，每一笔交易，银行都有记录并且能鉴别真伪，可作为客户授信评级的重要依据。中信银行采用大数据方案，可以结合实时、历史数据进行全局分析，风险管理部门现在可以每天评估客户的行为，并决定对客户的信用额度在同一天进行调整，原有内部系统、模型整体性能显著提高。

4. 大数据在零售行业的应用

零售行业大数据应用有两个层面：一个层面是零售行业可以了解客户的消费喜好和趋势，进行商品的精准营销，降低营销成本；另一个层面是依据客户购买的产品，为客户提供可能购买的其他产品，扩大销售额，也属于精准营销范畴。另外，零售行业还可以通过大数据掌握未来消费趋势，有利于热销商品的进货管理和过季商品的处理。零售行业的数据对于产品生产厂家是非常宝贵的，零售商的数据信息将会有助于资源的有效利用，降低产能过剩，厂商依据零售商的信息按实际需求进行生产，减少不必要的生产浪费。

未来考验零售企业的不再只是供应关系的好坏，而是要看挖掘消费者需求，以及高效整合供应链满足其需求的能力，因此信息科技水平的高低成为获得竞争优势的关键要素。不论是国际零售巨头，还是本土零售品牌，要想顶住日渐微薄的利润率带来的压力，在这片领域立于不败之地，就必须思考如何利用大数据为顾客带来更好的消费体验。

5. 大数据在农业的应用

大数据在农业的应用主要是指依据未来的商业需求预测来进行农牧产品的生产，降低菜贱伤农的概率。同时大数据的分析将会更加精确预测未来的天气气候，帮助农牧民做好自然灾害的预防工作。大数据同时也会帮助农民依据消费者的消费习惯来决定增加哪些品种的种植，减少哪些品种的生产，提高单位种植面积的产值，同时有助于快速销售农产品，完成资金回流。

由于农产品不容易保存，因此合理种植和养殖就显得十分重要。如果没有规划好，容易出现鸡蛋过剩、苹果过剩、大蒜过剩、莲藕过剩等伤农事件。借助于大数据提供的消费趋势报告和消费习惯报告，政府将为农牧业生产提供合理引导，建议依据需求进行生产，避免产能过剩造成的不必要资源和社会财富浪费。农业关乎到国计民生，科学的规划将有助于社会整体效率的提升，大数据技术可以帮助政府实现农业的精细化管理，实现科学决策。

6. 大数据在交通行业的应用

近年来，我国的智能交通已实现了快速发展，许多技术手段都达到了国际领先水平。但是，问题和困境也非常突出，从各个城市的发展状况来看，智能交通的潜在价值还没有得到有效挖掘。对交通信息的感知和收集有限，对存在于各个管理系统中的海量数据无法共享运用、有效分析，对交通态势的研判预测乏力，对公众的交通信息服务很难满足需求。这其中很重要的问题是对于海量数据尤其是半结构、非结构数据无能为力。

目前，交通的大数据应用主要在两个方面：一方面可以利用大数据传感器数据来了解车辆通行密度，合理进行道路规划包括单行线路规划；另一方面可以利用大数据来实现即时信号灯调度，提高已有线路通行能力。

7. 大数据在教育行业的应用

教育行业中的考试数据、学籍数据、教师数据、事业数据、经费数据、人口数据、研究数据等都分散在不同的机构和政府部门，很难形成大数据，这是需要统筹考虑解决的问题。

教育中有两个特定的领域会用到大数据：教育数据挖掘和学习分析。教育数据挖掘应用统计学、机器学习和数据挖掘的技术和开发方法，对教学和学习过程中收集的数据进行分析，检验学习理论并引导教育实践。学习分析应用信息科学、社会学、心理学、统计学、机器学习和数据挖掘的技术来分析从教育管理和服务过程中收集的数据，学习分析创建的应用程序直接影响教育实践。

8. 大数据在政府机构的应用

政府利用大数据技术可以了解各地区的经济发展情况、各产业发展情况、消费支出和产品销售情况，依据数据分析结果科学地制定宏观政策，平衡各产业发展，避免产能过剩，有效利用自然资源和社会资源，提高社会生产效率。在以下几个方面，可以进一步协助发挥政府机构的职能作用：

（1）重视应用大数据技术，盘活各地云计算中心资产，把原来大规模投资产业园、物联网产业园的政绩工程改造成智慧工程。

（2）在安防领域，应用大数据技术，提高应急处置能力和安全防范能力。

（3）在民生领域，应用大数据技术，提升服务能力和运作效率，以及个性化的服务，比如医疗、卫生、教育等部门。

（4）解决在金融、电信等领域数据分析的问题，提高国家的金融、电信安全水平，预防电信诈骗。

1.1.4 大数据的机遇与挑战

随着近年来大数据的不断升温，人们也逐渐意识到大数据中提到的数据是全部数据，而不是随机采样；预测是大体方向，而不是精确制导。随着对大数据研究的不断深入，人们越来越意识到对数据的利用可以为其生产生活带来巨大的便利，同时也带来了不小的挑战。

1. 大数据的安全与隐私问题

在互联网上浏览网页，就会留下一连串的浏览痕迹；注册登录网站需要输入个人的重要信息，例如用户名、登录密码、手机号，甚至是身份证号、住址、银行卡等信息。通过相关的数据分析，就可以轻易挖掘出人们的行为习惯和个人重要信息。如果这些信息运用得当，可以帮助相关领域的企业随时了解客户的需求和习惯，便于企业调整相应的产品生产计划，取得更大的经济效益。但若是这些重要的信息被不良分子窃取，随之而来的就是个人信息泄露、财产丢失等安全性问题。

2. 对现有技术的挑战

（1）对现有数据库管理技术的挑战。

传统的数据库部署不能处理 TB 级别的数据，也不能很好地支持高级别的数据分析。急速膨胀的数据体量即将超越传统数据库的管理能力。如何构建全球级的分布式数据库，可以扩展到数百万的机器，数以百计的数据中心，上万亿行的数据，是今后大数据处理需要解决的问题。

（2）对经典数据库技术的挑战。

经典数据库并没有考虑数据的多类别，SQL（结构化数据查询语言）在设计的一开始是没有考虑非结构化数据的。

（3）实时性的技术挑战。

传统的数据仓库系统和各类 BI（Business Intelligence）应用对处理时间的要求并不高，因此这类应用往往运行一两天获得结果依然是可行的。但实时处理的要求是区别大数据应用和传统数据仓库技术、BI 技术的关键差别之一。

（4）对网络架构、数据中心、运维的挑战。

人们每天创建的数据量正呈爆炸式增长，但就数据保存来说，我们的技术改进不大，而数据丢失的可能性却不断增加。如此庞大的数据量首先在存储上就会是一个非常严重的问题，硬件的更新速度将是大数据发展的基石。

1.2 大数据处理技术

面对大数据的全新特征，既有的技术架构和路线已经无法高效地处理如此海量的数据，而对于相关组织来说，如果投入巨大采集的数据无法及时处理反馈有效信息，那将是得不偿失的。可以说，大数据时代对人类的数据驾驭能力提出了新的挑战，也为人们获得更为深刻、全面的洞察能力提供了前所未有的空间与潜力。

大数据技术的不同技术层面功能和计算模式如表 1-1 和表 1-2 所示。

表 1-1 大数据技术层面功能

技术层面	功能
数据采集	利用 ETL 工具将分布的异构数据源中的数据如关系数据、平面数据文件等，抽取到临时中间层后进行清洗、转换、集成，最后加载到数据仓库或数据集市中，成为联机分析处理、数据挖掘的基础；或者也可以把实时采集的数据作为流计算系统的输入，进行实时处理分析
数据存储和管理	利用分布式文件系统、数据仓库、关系数据库、NoSQL 数据库、云数据库等，实现对结构化、半结构化和非结构化海量数据的存储和管理
数据处理与分析	利用分布式并行编程模型和计算框架，结合机器学习和数据挖掘算法，实现对海量数据的处理和分析；对分析结果进行可视化呈现，帮助人们更好地理解数据、分析数据
数据隐私和安全	在从大数据中挖掘潜在的巨大商业价值和学术价值的同时，构建隐私数据保护体系和数据安全体系，有效保护个人隐私和数据安全

表 1-2 大数据计算模式

计算模式	解决问题	代表产品
批处理计算	针对大规模数据的批量处理	MapReduce、Spark 等
流计算	针对流数据的实时计算	Storm、S4、Flume、Streams、Puma、DStream、Super Mario、银河流数据处理平台等
图计算	针对大规模图结构数据的处理	Pregel、GraphX、Giraph、PowerGraph、Hama、GoldenOrb 等
查询分析计算	大规模数据的存储管理和查询分析	Dremel、Hive、Cassandra、Impala 等

1.3 大数据与云计算

云计算（Cloud Computing）是基于互联网的相关服务的增加、使用和交付模式，通常涉及通过互联网来提供动态易扩展且经常是虚拟化的资源。云是网络、互联网的一种比喻说法，过去往往用云来表示网络，后来也用来表示互联网和底层基础设施的抽象。狭义云计算指 IT 基础设施的交付和使用模式，指通过网络以按需、易扩展的方式获得所需资源；广义云计算指服务的交付和使用模式，指通过网络以按需、易扩展的方式获得所需服务。这种服务可以是 IT 和软件、互联网相关，也可以是其他服务。它意味着计算能力也可以作为一种商品通过互联网进行流通。

大数据，或称海量数据，指的是所涉及的资料量规模巨大到无法通过目前的主流软件工具，在合理时间内达到撷取、管理、处理并整理成为帮助政府、企业经营决策的依据。

从技术上看，大数据与云计算的关系就像一枚硬币的正反面一样密不可分。大数据必然无法用单台的计算机进行处理，必须采用分布式计算架构。它的特色在于对海量数据的挖掘，但它必须依托云计算的分布式处理、分布式数据库、云存储和虚拟化技术。云计算和大数据的关系如图 1-1 所示。

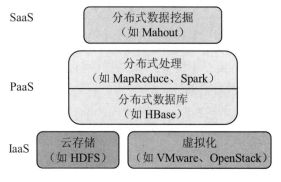

图 1-1 云计算和大数据的关系图

简单来说，云计算是硬件资源的虚拟化，而大数据是海量数据的高效处理。虽然从这个解释来看也不是完全贴切，但是却可以帮助对这两个名字不太明白的人很快理解其区别。当然，如果解释更形象一点的话，云计算相当于我们的计算机和操作系统将大量的硬件资源虚拟化后再进行分配使用。

可以说，大数据相当于海量数据的"数据库"，通观大数据领域的发展我们也可以看出，当前的大数据发展一直在向着近似于传统数据库体验的方向发展，一句话就是，传统数据库给大数据的发展提供了足够大的空间。

大数据的总体架构包括三层：数据存储、数据处理和数据分析。数据先要通过存储层存储下来，然后根据数据需求和目标来建立相应的数据模型和数据分析指标体系对数据进行分析产生价值，而中间的时效性又通过中间数据处理层提供的强大的并行计算和分布式计算能力来完成。三者相互配合，这让大数据产生最终价值。

云计算未来的趋势是：云计算作为计算资源的底层，支撑着上层的大数据处理；大数据的发展趋势是，实时交互式的查询效率和分析能力。

1.4 本章小结

大数据包含庞杂的知识体系，在具体学习相关技术之前，有必要对其有清晰直观的认识。大数据具有规模性、多样性、价值密度和高速性的特征。它虽然在金融行业、互联网、医疗行业、社交网络等方面改变着人们的生活，但是也对人们的信息安全和现有技术提出了挑战。大数据技术主要包含数据采集、数据存储和管理、数据处理与分析、数据隐私和安全等层面，而常用的计算模式有批处理计算、流计算和图计算等。本章最后阐述了大数据与云计算之间的区别和联系，使读者对两者有个清楚的了解。

第 2 章　大数据处理平台 Hadoop

　　Apache Hadoop 是一款支持数据密集型分布式应用并以 Apache 2.0 许可协议发布的开源软件框架，支持在商品硬件构建的大型集群上运行应用程序。Hadoop 是根据 Google 公司发表的 MapReduce 和 GFS 论文自行开发而成的。

　　Hadoop 框架透明地为应用提供可靠性和数据移动，它实现了名为 MapReduce 的编程范式：应用程序被切分成许多小部分，而每个部分都能在集群中的任意节点上执行或重新执行。此外，Hadoop 还提供了分布式文件系统，用以存储所有计算节点的数据，这为整个集群带来了非常高的带宽。MapReduce 和分布式文件系统的设计，使得整个框架能够自动处理节点故障。通过本章学习，您将会掌握 Hadoop 生态系统和版本变迁等知识。

2.1　Hadoop 生态系统

　　Hadoop 是一个能够对大量数据进行分布式处理的软件框架，具有可靠、高效、可伸缩的特点。Hadoop 2.0 版本引入了 HA（High Avalability，高可用性）和 Yarn（资源调度），这是与 Hadoop 1.0 的最大区别。Hadoop 1.0 生态系统如图 2-1 所示。

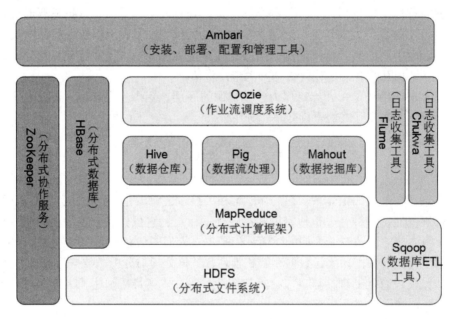

图 2-1　Hadoop 1.0 生态系统

　　Hadoop 2.0 主要由三部分组成：HDFS 分布式文件系统、MapReduce 编程模型和 Yarn 资源管理。Hadoop 2.0 生态系统如图 2-2 所示。

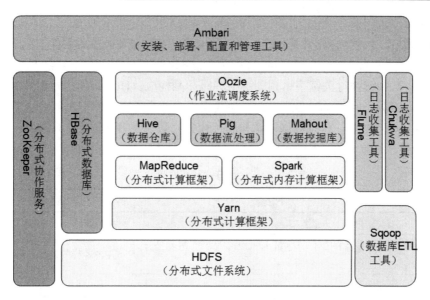

图 2-2　Hadoop 2.0 生态系统

由图 2-1 和图 2-2 可以看出，整个 Hadoop 家族由以下几个子项目组成：

（1）HDFS。

对于分布式计算，每个服务器必须具备对数据的访问能力，这就是 HDFS（Hadoop Distributed File System）所起到的作用。在处理大数据的过程中，当 Hadoop 集群中的服务器出现错误时，整个计算过程并不会终止，同时 HDFS 可以保障在整个集群中发生故障错误时的数据冗余。当计算完成时将结果写入 HDFS 的一个节点之中，HDFS 对存储的数据格式并无苛刻的要求，数据可以是非结构化或其他类别的，而关系数据库在存储数据之前需要将数据结构化并定义 Schema。

（2）MapReduce。

MapReduce 是一个计算模型，用于大规模数据集的并行运算。它极大地方便了编程人员在不会分布式并行编程的情况下，将自己的程序运行在分布式系统上。MapReduce 的重要创新是当处理一个大数据集查询时会将其任务分解并在运行的多个节点中处理。当数据量很大时就无法在一台服务器上解决问题，此时分布式计算的优势就体现出来了，将这种技术与 Linux 服务器结合可以获得性价比极高的替代大规模计算阵列的方法。Hadoop MapReduce 级的编程利用 Java APIs，并可以手动加载数据文件到 HDFS 中。

（3）ZooKeeper。

ZooKeeper 是一个分布式应用程序协调服务，是 Hadoop 和 HBase 的重要组件。它是一个为分布式应用提供一致性服务的软件，提供的功能包括配置维护、域名服务、分布式同步、组服务等。

ZooKeeper 集群提供了 HA，可以保证在其中一些机器死机的情况下仍可以提供服务，而且数据不会丢失；所有 ZooKeeper 服务的数据都存储在内存中，且数据都是副本。ZooKeeper 集群中包括领导者（leader）和跟随者（follower）两种角色，当客户端进行读取时，跟随者的服务器负责给客户端响应；客户端的所有更新操作都必须通过领导者来处理。当更新被大多数 ZooKeeper 服务成员持久化后，领导者会给客户端响应。

(4) HBase。

HBase 是一个针对结构化数据的可伸缩、高可靠、高性能、分布式和面向列的动态模式数据库。与传统关系数据库不同，HBase 采用了 BigTable 的数据模型：增强的稀疏排序映射表（key/value），其中键由行关键字、列关键字和时间戳构成。HBase 提供了对大规模数据的随机、实时读写访问，使用 Hadoop HDFS 作为其文件存储系统，同时 HBase 中保存的数据可以使用 MapReduce 来处理，它将数据存储和并行计算完美地结合在一起。HBase 与关系数据库的对比如表 2-1 所示。

表 2-1　HBase 与关系数据库的对比

对比项	HBase	关系数据库
数据类型	HBase 只有 bytes 类型	拥有丰富的数据类型和存储方式
数据操作	只有很简单的操作，如插入、删除等，表与表是分离的，之间没有复杂的关系	各种各样的连接操作和函数
数据维护	更新操作时，会将原有数据保留，所以它实际上是插入了新数据	直接修改原数据
存储方式	基于列存储的，每个列族都有自己的文件，不同的列族是分开的	基于表结构和行来存储
可扩展性	支持随意的扩展，而不需要改变表内原有的数据	修改表结构需要复杂的操作
事务	没有复杂的事务支持，只有简单的行级事务	ACID 保证
索引	没有二级索引	拥有丰富的索引支持

(5) Hive。

Hive 是基于 Hadoop 的一个数据仓库工具，由 Facebook 开源，最初用于海量结构化日志数据统计，可以将结构化的数据文件映射为一张数据库表，并提供简单的 SQL 查询功能，可以将 SQL 语句转换为 MapReduce 任务运行。通常用于进行离线数据处理（采用 MapReduce），可以认为是一个从 HQL（Hive QL）到 MapReduce 的语言翻译器。

Hive 的特点如下：

- 可扩展。Hive 可以自由地扩展集群的规模，一般情况下不需要重启服务。
- 支持 UDF。Hive 支持用户自定义函数，用户可以根据自己的需要来实现。
- 容错。良好的容错性，节点失效时 SQL 依然可以正确执行到结束。
- 自由的定义输入格式。默认 Hive 支持 txt、rc、sequence 等，用户可以自由地定制自己想要的输入格式。
- 可以根据字段创建分区表，如根据日志数据中的日期。

(6) Pig。

Pig 是一个高级过程语言，它简化了 Hadoop 常见的工作任务，适合于使用 Hadoop 和 MapReduce 平台来查询大型半结构化数据集。通过允许对分布式数据集进行类似 SQL 的查询，Pig 可以简化 Hadoop 的使用。Pig 可以加载数据、表达转换数据和存储最终结果。Pig 内置的操作使得半结构化数据变得有意义（如日志文件），同时 Pig 可以扩展使用 Java 中添加的自定义数据类型并支持数据转换。

可以避免用户书写 MapReduce 程序，由 Pig 自动转成。任务编码的方式允许系统自动去优化执行过程，从而使用户能够专注于业务逻辑，用户可以轻松地编写自己的函数来进行特殊用途的处理。

（7）Mahout。

Mahout 起源于 2008 年，最初是 Apache Lucent 的子项目，它在极短的时间内取得了长足的发展，现在是 Apache 的顶级项目。Mahout 的主要目标是创建一些可扩展的机器学习领域经典算法的实现，旨在帮助开发人员更加方便快捷地创建智能应用程序。Mahout 现在已经包含了聚类、分类、推荐引擎（协同过滤）和频繁集挖掘等广泛使用的数据挖掘方法。除了算法，Mahout 还包含数据的输入/输出工具、与其他存储系统（如数据库、MongoDB 或 Cassandra）集成等数据挖掘支持架构。

（8）Sqoop。

Sqoop 是 Hadoop 与结构化数据存储互相转换的开源工具。可以使用 Sqoop 从外部的数据存储将数据导入到 Hadoop 分布式文件系统或相关系统，如 Hive 和 HBase。Sqoop 也可以用于从 Hadoop 中提取数据，并将其导出到外部的数据存储（如关系数据库和企业数据仓库），如 MySQL、Oracle、SQL Server，还可以通过脚本快速地实现数据的导入/导出。

（9）Flume。

Flume 是 Cloudera 提供的一个高可用、高可靠、分布式的海量日志采集、聚合和传输的系统。Flume 支持在日志系统中定制各类数据发送方，用于收集数据。同时，Flume 提供对数据进行简单处理，并写到各种数据接收方（可定制）的能力。它将数据从产生、传输、处理并最终写入目标路径的过程抽象为数据流，在具体的数据流中，数据源支持在 Flume 中定制数据发送方，从而支持收集各种不同协议数据。同时，Flume 数据流提供对日志数据进行简单处理的能力，如过滤、格式转换等。总的来说，Flume 是一个可扩展、适合复杂环境的海量日志收集系统。

（10）Chukwa。

Chukwa 是一个开源的用于监控大型分布式系统的数据收集系统，构建在 Hadoop 的 HDFS 和 MapReduce 框架之上。Chukwa 还包含了一个强大和灵活的工具集，可以用于展示、监控和分析已收集的数据。

（11）Oozie。

Oozie 是一个工作流引擎服务器，用于运行 Hadoop MapReduce 任务工作流（包括 MapReduce、Pig、Hive、Sqoop 等）。Oozie 工作流通过 HPDL（Hadoop Process Defintion Langue）来构建，工作流定义通过 HTTP 提交，可以根据目录中是否有数据来运行任务，任务之间的依赖关系通过工作流来配置，任务可以定时调度。JobConf 类要配置的内容，通过在工作流（XML 文件）中定义，实现了配置与代码的分离。

（12）Ambari。

Ambari 是一种基于 Web 的工具，用于创建、管理、监视 Hadoop 的集群，支持 Hadoop HDFS、Hadoop MapReduce、Hive、Hcatalog、HBase、ZooKeeper、Oozie、Pig 和 Sqoop 等的集中管理。

2.2　Hadoop 架构

Hadoop 是一个存储和处理大规模数据的开源软件框架，实现在大量计算机组成的集群中对海量数据进行分布式存储计算。Hadoop 最初由 Doug Cutting 根据 Google 的 GFS 和 MapReduce 思想，采用 Java 语言开发而创建。

由于 Hadoop 采用了分布式存储方式和 Java 语言开发，这使得 Hadoop 可以部署在不同操作系统平台和通用的计算机集群中。Hadoop 中 HDFS 的数据管理能力能够快速高效地读写文件，同时还采用了存储冗余数据的方式保证了数据的安全性。MapReduce 处理分布式任务时的简单方便，以及开源特性，使 Hadoop 在大数据领域被广泛使用。

用户可以轻松地在 Hadoop 上开发和运行处理海量数据的分布式应用程序，Hadoop 主要有以下几个优点：

（1）高扩展性。Hadoop 支持集群节点的动态增加，方便集群的扩展，可以根据需要扩展到上千节点。

（2）高容错性。Hadoop 能够自动保存数据的多个副本，当有集群节点数据丢失时，能够自动实现数据的备份。

（3）高效性。Hadoop 可以充分地利用网络、磁盘 I/O 资源和备份数据实现文件的快速读写。

（4）高可靠性。正在执行的子任务由于某种原因挂掉的话，Hadoop 可以自动让其他节点继续执行该子任务，不影响整个任务的执行。

（5）低成本。Hadoop 集群不用部署在专用服务器上，只需部署在通用的计算机中即可；而且 Hadoop 生态系统绝大部分都是开源的，用户很容易拿来开发使用。

HDFS 和 MapReduce 是 Hadoop 的两大核心，整个 Hadoop 的体系结构主要是 HDFS 为海量数据提供了分布式存储，MapReduce 为海量数据的规则提取提供了并行计算。

2.2.1 HDFS

HDFS 是一个分布式文件系统，采用了主从（Master-Slave）架构模型。一个 HDFS 集群是由一个 NameNode（Master）节点和若干 DataNode（Slave）节点组成。其中 NameNode 节点作为主服务器，管理整个文件系统的命名空间和客户端对文件的访问操作，比如打开、关闭、重命名文件或目录等，同时它也负责数据块（在 HDFS 中，一个文件被分成若干数据块，这些数据块存放在 DataNode 上）到具体 DataNode 的映射。DataNode 管理存储的数据，处理客户端的文件读写请求，并在 NameNode 的统一调度下进行数据块的创建、删除和复制工作。

为了避免数据的丢失，HDFS 默认采用了三个冗余备份，并将这些数据放在不同的 DataNode 上。

2.2.2 MapReduce

Hadoop MapReduce 是 Doug Cutting 受到 Google 发表的关于 MapReduce 的论文的启发而开发出来的。MapReduce 是一个简单易用的并行计算模型，它将运行于大规模集群上的复杂的并行计算过程高度地抽象为两个函数：Map 和 Reduce。MapReduce 应用程序能够以一种可靠容错的方式并行处理大数据，实现了在 Hadoop 集群上数据和任务的并行计算与处理。

MapReduce 应用通常会把输入文件切分为若干数据块，每个数据块会启动一个相应的 Map 来并行处理，然后把 Map 处理后的结果输入给 Reduce 来进行汇总。整个框架负责任务的调度和监控，以及重新执行已经失败的任务。

通常，MapReduce 框架和分布式文件系统是运行在一组相同的节点上的，也就是说，计算节点和存储节点在一起。这种做法可以让存储节点上运行 MapReduce 程序，减少集群内部数据传输量，提高运行效率。

2.2.3 Yarn

Yarn 是 Hadoop 2.0 版本中的一个新增特性，主要负责集群的资源管理和调度。Yarn 不仅可以支持 MapReduce 计算，还支持 Hive、HBase、Pig、Spark 等应用，这样就可以方便地使用 Yarn 从系统层面对集群进行统一的管理。

Yarn 仍然采用 Master-Slave 结构，在整个资源管理框架中，ResourceManager 为 Master，NodeManager 为 Slave，ResourceManager 负责整个系统的资源管理和分配，NodeManager 负责每个 DataNode 上的资源和任务管理。

2.3 Hadoop 版本变迁

2003 年至 2004 年，Google 公布了部分 GFS 和 MapReduce 思想的细节，受此启发的 Doug Cutting 等人用两年的业余时间实现了 DFS 和 MapReduce 机制，使 Nutch 性能飙升，然后 Doug Gutting 加入 Yahoo!。2005 年，Hadoop 作为 Lucene 的子项目 Nutch 的一部分正式被引入 Apache 基金会。2006 年 2 月被分离出来，成为一套完整独立的软件，起名为 Hadoop。

Hadoop 这个名字不是一个缩写，而是一个生造出来的词，是 Hadoop 之父 Doug Cutting 以儿子的毛绒玩具象命名的。总结起来，Hadoop 起源于 Google 的三大论文：GFS（Google 的分布式文件系统 Google File System）、MapReduce（Google 的 MapReduce 开源分布式并行计算框架）和 BigTable（一个大型的分布式数据库）。最终 GFS 和 MapReduce 演变成 Hadoop 的 HDFS 和 MapReduce，BigTable 则成为了 HBase。

2.3.1 Hadoop 发展史

2004 年：最初的版本（现在称为 HDFS 和 MapReduce）由 Doug Cutting 和 Mike Cafarella 开始实施。

2005 年 12 月：Hadoop 由 Lucene 改名为 Nutch，并被移植到新的框架，Hadoop 在 20 个节点上稳定运行。

2006 年 1 月：Doug Cutting 加入 Yahoo!。

2006 年 2 月：Apache Hadoop 项目正式启动，以支持 MapReduce 和 HDFS 的独立发展。

2008 年 9 月：Hive 成为 Hadoop 的子项目。

2009 年 3 月：Cloudera 推出 CDH（Cloudera's Distribution Including Apache Hadoop），相对于 Apache Hadoop，其具有版本层次更加清晰、系统稳定性好的特点，且它提供了适用于各种操作系统的 Hadoop 安装包，可以直接使用 apt-get 或 yum 命令进行安装，更加省事。

2009 年 7 月：MapReduce 和 HDFS 成为 Hadoop 项目的独立子项目，Avro 和 Chukwa 成为 Hadoop 新的子项目。

2.3.2 如何选择 Hadoop 开发版本

Hadoop 版本比较混乱，让很多用户不知所措。实际上，截至 2017 年 12 月 1 日，Hadoop 有两系列版本：Hadoop 1.x 和 Hadoop 2.x。Hadoop 1.0 由一个分布式文件系统 HDFS 和一个离线计算框架 MapReduce 组成，而 Hadoop 2.x 包含一个支持 NameNode 横向扩展的 HDFS、一个资源管理系统 Yarn 和一个运行在 Yarn 上的离线计算框架 MapReduce。相比于 Hadoop 1.x，Hadoop 2.x 功能更加强大，且具有更好的扩展性，并支持多种计算框架。

当我们决定是否采用某个软件用于开源环境时，通常需要考虑以下几个因素：

（1）是否为开源软件，即是否免费。

（2）是否有稳定版，这个一般软件官方网站会给出说明。

（3）是否经实践验证，这个可通过查询是否有一些知名的公司已经在生产环境中使用来了解。

（4）是否有强大的社区支持，当出现一个问题时，能够通过社区、论坛等网络资源快速获取解决方法。

考虑到以上几个因素，我们采用 Hadoop 2.6 开发学习。Hadoop 2.6 目前版本稳定，广泛应用于生产环境，而且 Spark 也有对该平台的开发版本，方便后续扩充。针对 Hadoop 1.x 版本中容易出现的单点故障问题，在 Hadoop 2.x 中可以通过 ZooKeeper 配置高可用集群，集群中允许有多个 NameNode（生成环境下一般不超过三个），其中一个处于 Active 状态，其余的处于 Standby 状态。当一个 NameNode 所在的服务器死机时，可以在数据不丢失的情况下，集群自动推选一个处于 Standby 状态的 NameNode 充当 Master，不影响整个集群的使用。

2.4 本章小结

Hadoop 目前已经在各个领域得到了广泛应用，如淘宝、百度和 Facebook 等。本章首先介绍了 Hadoop 1.0 和 Hadoop 2.0 生态系统，并对 ZooKeeper、HBase、Hive、Pig、Mahout、Sqoop、Flume、Chukwa、Oozie 和 Ambari 子模块进行了简要概述；然后介绍了其三大组成部分：HDFS、MapReduce 和 Yarn；最后介绍了 Hadoop 版本发展史，并针对开发人员如何选择给出了相关建议。

第 3 章　Hadoop 平台搭建

　　Hadoop 是一个开源的、可运行于 Linux 集群上的分布式计算平台，借助于 Hadoop，用户可以轻松地保存大数据并对其进行数据分析。工欲善其事必先利其器，在学习 Hadoop 之前首先将 Hadoop 集群搭建起来。本章将会介绍基础环境的配置、Hadoop 集群的搭建、验证以及搭建注意事项。

3.1　基础环境配置

　　针对 Hadoop 学习，本书采用 VMware9+CentOS7 64 位方案，其中主机内存应不小于 8GB 且操作系统为 64 位，一个 Linux 主节点部署 CentOS7 桌面版，并为其分配 2GB 内存，两个 Linux 从节点部署 CentOS mini 版，分配 1GB 内存，具体安装参考附录。为了方便操作和文件传输，建议安装 Xshell 和 Xftp。

　　CentOS mini 版本没有 ifconfig 等命令，查看 IP 需要使用命令 ip addr，由于没有 ifconfig 等命令不方便操作，可以安装 net-tools，下载地址为 https://centos.pkgs.org/7/centos-x86_64/net-tools-2.0-0.17.20131004git.el7.x86_64.rpm.html。将 net-tools 上传到/soft 文件夹下（本书所用到的 Linux 安装软件均在/soft 目录下），并使用如下命令安装：

rpm -ivh net-tools-2.0-0.17.20131004git.el7.x86_64.rpm

　　这样就可以使用网络常用命令。在 CentOS7 中已经默认安装 SSH，此步骤可以省略，如果是其他 Linux 版本，如 Ubuntu，需要安装 SSH，可以在https://pkgs.org/download/openssh下载编译好的安装包直接安装，也可以通过 yum 或 apt 等命令在线安装。

3.2　Hadoop 配置文件修改

1. 配置静态 IP

　　通过 dhclient 命令获得动态 IP，这样可以避免 IP 冲突并获得网络相关信息。修改 IP 地址命令为：vi /etc/sysconfig/network-scripts/ifcfg-ens33，内容修改为如下：

```
BOOTPROTO=static            #dhcp 改为 static
ONBOOT=yes                  #开机启用本配置
IPADDR=192.168.254.128      #静态 IP
GATEWAY=192.168.254.2       #默认网关
NETMASK=255.255.255.0       #子网掩码
DNS1=114.114.114.114        #DNS 配置
NM_CONTROLLED=no
```

　　默认网关可以通过 route 命令查看设置，每台虚拟机对应一个静态 IP。配置完毕后执行命令 service network restart 重启服务，或者重启计算机，如果进入 CentOS 很慢，说明网络配置有误，需要重新更正。对三个节点均需进行配置，其中主节点（桌面版）IP 为 192.168.254.128。

两个从节点（非桌面版）IP 分别为 192.168.254.129 和 192.168.254.131。

执行 ping 命令，如 ping 192.168.254.129，有正确回复表示网络配置正常，如图 3-1 所示。

```
[root@master data]# ping 192.168.254.129
PING 192.168.254.129 (192.168.254.129) 56(84) bytes of data.
64 bytes from 192.168.254.129: icmp_seq=1 ttl=64 time=23.5 ms
64 bytes from 192.168.254.129: icmp_seq=2 ttl=64 time=1.37 ms
64 bytes from 192.168.254.129: icmp_seq=3 ttl=64 time=2.17 ms
64 bytes from 192.168.254.129: icmp_seq=4 ttl=64 time=1.18 ms
64 bytes from 192.168.254.129: icmp_seq=5 ttl=64 time=1.12 ms
```

图 3-1　master 节点 ping 192.168.254.129 节点

2．配置/etc/hostname

该文件是配置主机名，每台虚拟机各不相同。IP 为 128 的设置为 master（主节点），129 的为 slave1（从节点 1），131 的为 slave2（从节点 2）。执行 ping 命令，如 ping slave1，有正确回复表示配置正常，如图 3-2 所示。

```
[root@master data]# ping slave1
PING slave1 (192.168.254.129) 56(84) bytes of data.
64 bytes from slave1 (192.168.254.129): icmp_seq=1 ttl=64 time=1.40 ms
64 bytes from slave1 (192.168.254.129): icmp_seq=2 ttl=64 time=1.37 ms
64 bytes from slave1 (192.168.254.129): icmp_seq=3 ttl=64 time=1.46 ms
64 bytes from slave1 (192.168.254.129): icmp_seq=4 ttl=64 time=2.20 ms
```

图 3-2　master 节点 ping slave1 节点

3．配置/etc/hosts

该文件是配置 IP 和主机名的对应关系，三个虚拟机均采用如下配置（第一行配置不能丢掉，否则 Hadoop 启动的时候会报错）：

```
127.0.0.1   localhost
192.168.254.128   master
192.168.254.129   slave1
192.168.254.131   slave2
```

可以在任意主机上执行“ping 主机名”命令，确定是否有正确回复。

4．SSH 免密钥登录

Hadoop 中的 NameNode 和 DataNode 数据通信采用了 SSH 协议，需要配置 master 对各 slave 节点的免密钥登录。

在 slave1 和 slave2 中执行命令 mkdir /root/.ssh。

在 master 节点执行如下操作：

```
ssh-keygen -t dsa                    #采用了 dsa 加密，也可以采用 rsa 加密
cat /root/.ssh/id_dsa.pub >> /root/.ssh/authorized_keys
scp /root/.ssh/authorized_keys   slave1:/root/.ssh/
scp /root/.ssh/authorized_keys   slave2:/root/.ssh/
```

配置完成后，在 master 节点可以通过 ssh slave1 免密钥登录 slave1，如果还需要密码，说明上面操作有误（容易犯的错误是误把/root/.ssh 认为是一个文件）。

5．安装 JDK

下载 JDK 的网址为http://www.oracle.com/technetwork/java/javase/downloads/index.html，为了方便 Scala 和 Spark 的学习，本书采用 JDK1.8 安装，如图 3-3 所示。

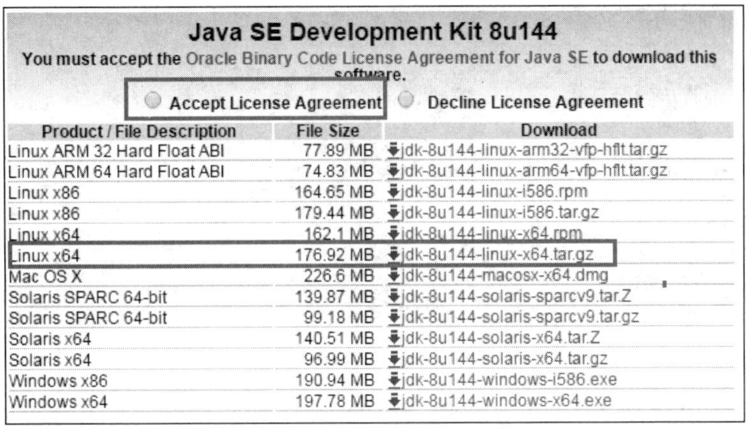

Java SE Development Kit 8u144

You must accept the Oracle Binary Code License Agreement for Java SE to download this software.

Accept License Agreement　Decline License Agreement

Product / File Description	File Size	Download
Linux ARM 32 Hard Float ABI	77.89 MB	jdk-8u144-linux-arm32-vfp-hflt.tar.gz
Linux ARM 64 Hard Float ABI	74.83 MB	jdk-8u144-linux-arm64-vfp-hflt.tar.gz
Linux x86	164.65 MB	jdk-8u144-linux-i586.rpm
Linux x86	179.44 MB	jdk-8u144-linux-i586.tar.gz
Linux x64	162.1 MB	jdk-8u144-linux-x64.rpm
Linux x64	176.92 MB	jdk-8u144-linux-x64.tar.gz
Mac OS X	226.6 MB	jdk-8u144-macosx-x64.dmg
Solaris SPARC 64-bit	139.87 MB	jdk-8u144-solaris-sparcv9.tar.Z
Solaris SPARC 64-bit	99.18 MB	jdk-8u144-solaris-sparcv9.tar.gz
Solaris x64	140.51 MB	jdk-8u144-solaris-x64.tar.Z
Solaris x64	96.99 MB	jdk-8u144-solaris-x64.tar.gz
Windows x86	190.94 MB	jdk-8u144-windows-i586.exe
Windows x64	197.78 MB	jdk-8u144-windows-x64.exe

图 3-3　JDK 下载

下载完毕后，执行如下操作：

```
mkdir /Java
tar -zxvf jdk-8u144-linux-x64.tar.gz -C /Java
scp -r /Java/   slave1:/
scp -r /Java/   slave2:/
```

三个虚拟机修改环境变量（vi /etc/profile），在文件末尾添加：

```
export JAVA_HOME=/Java/jdk1.8.0_144
export PATH=$PATH:$JAVA_HOME/bin:$JAVA_HOME/jre/bin
export CLASSPATH=$CLASSPATH:.:$JAVA_HOME/lib:$JAVA_HOME/jre/lib
```

执行命令 source /etc/profile，输入 java -version，会有如下提示：

```
java version "1.7.0_75"
OpenJDK Runtime Environment (rhel-2.5.4.2.el7_0-x86_64 u75-b13)
OpenJDK 64-Bit Server VM (build 24.75-b04, mixed mode)
```

在 Linux 中环境变量可以配置在/etc/profile 中，也可以配置在~/.bashrc 下，区别是前者对所有用户都生效，而后者只对当前用户生效。

6. 关闭所有虚拟机防火墙

```
systemctl stop firewalld.service        #停止 firewall
systemctl disable firewalld.service     #禁止 firewall 开机启动
```

7. 配置时钟同步

在https://pkgs.org/download/ntp上下载 autogen-libopts-5.18-5.el7.x86_64.rpm、ntpdate-4.2.6p5-25.el7.centos.x86_64.rpm 和 ntp-4.2.6p5-25.el7.centos.x86_64.rpm 三个包。

（1）在三个虚拟机上分别安装刚下载的三个包。

```
rpm -ivh autogen-libopts-5.18-5.el7.x86_64.rpm
rpm -ivh ntpdate-4.2.6p5-25.el7.centos.x86_64.rpm
rpm -ivh ntp-4.2.6p5-25.el7.centos.x86_64.rpm
```

（2）设置系统开机自动启动并启动服务。

```
systemctl enable ntpd
systemctl start ntpd
```

（3）在 master 节点修改/etc/ntp.conf 文件。

/etc/ntp.conf 文件内容修改如图 3-4 所示。

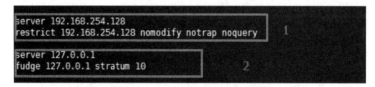

图 3-4 master 节点/etc/ntp.conf

第一处意思是 IP 地址从 192.168.254.1 到 192.168.254.254，默认网关为 255.255.255.0 的机器都可以从 NTP 服务器来同步时间；第二处指明谁作为 NTP 服务器；第三处是指当服务器与公用的时间服务器失去联系时以本地时间为客户端提供时间服务。

配置文件修改完成，重启服务 service ntpd restart。

（4）在 slave1 和 slave2 节点修改/etc/ntp.conf 文件。

各 slave 节点/etc/ntp.conf 文件内容修改如图 3-5 所示。

图 3-5 slave 节点/etc/ntp.conf

指定 192.168.254.128 作为 NTP 服务器。配置文件修改完成，重启服务 service ntpd restart。在 slave2 节点运行 ntpstat，可以看到如图 3-6 所示的提示。

图 3-6 ntpstat 启动信息

8. Hadoop 安装

Hadoop 版本采用的是 2.6.5，该版本是 Hadoop 的一个稳定版本，被广泛应用，下载地址为http://hadoop.apache.org/releases.html。

将其上传到 master 节点并解压到/hadoop 文件夹中，该文件夹存放本书所用到的软件，如 HBase、Hive、Scala、Spark 等。

　　　　tar -zxvf hadoop-2.6.5.tar.gz -C /hadoop/

进入/hadoop/hadoop-2.6.5/etc/hadoop/目录修改里面的配置文件。

（1）修改 hadoop-env.sh 和 yarn-env.sh。

在 Hadoop 中，以 env.sh 结尾的文件通常是配置所需的环境变量。

修改 Java 环境变量，如图 3-7 所示将其值改为前面 Java 的安装路径 export JAVA_HOME=/Java/jdk1.8.0_144/。

```
# The java implementation to use.
export JAVA_HOME=${JAVA_HOME}
```

图 3-7　JAVA_HOME 配置

（2）修改 core-site.xml。

core-site.xml 是 Hadoop 的全局配置文件，主要设置一些核心参数信息，如 fs.defaultFS 设置集群的 HDFS 访问路径，hadoop.tmp.dir 指定 NameNode、DataNode 等存放数据的公共目录。

首先创建如下目录：

  mkdir -p /hadoop/hdfs/tmp

然后在<configuration>里面添加如图 3-8 所示的内容。

```xml
<configuration>
    <property>
        <name>fs.defaultFS</name>
        <value>hdfs://master:9000</value>
    </property>
    <property>
        <name>hadoop.tmp.dir</name>
        <value>/hadoop/hdfs/tmp</value>
    </property>
</configuration>
```

图 3-8　core-site.xml 配置

（3）修改 hdfs-site.xml。

该文件是 HDFS 的配置文件，在<configuration>里面添加如图 3-9 所示的内容。

```xml
<configuration>
    <property>
        <name>dfs.replication</name>
        <value>2</value>
    </property>
</configuration>
```

图 3-9　hdfs-site.xml 配置

dfs.replication 用来设置副本存放个数，在实际生产中还会设置 dfs.namenode.name.dir 和 dfs.namenode.data.dir（NameNode 和 DataNode 的存放路径），同时将 core-site.xml 中的 hadoop.tmp.dir 配置去掉，避免冲突。

（4）修改 mapred-site.xml。

该文件是 MapReduce 的配置文件，由于 Hadoop 中不存在该文件，因此首先复制一个：

  cp /hadoop/hadoop-2.6.5/etc/hadoop/mapred-site.xml.template　/hadoop/hadoop-2.6.5/etc/hadoop/mapred-site.xml

然后将其修改为如图 3-10 所示的内容，指定由 Yarn 作为 MapReduce 的程序运行框架。如果没有配置这项，那么提交的程序只会运行在 local 模式，而不是分布式模式。

```
<configuration>
    <property>
        <name>mapreduce.framework.name</name>
        <value>yarn</value>
    </property>
</configuration>
```

图 3-10　mapred-site.xml 配置

（5）修改 yarn-site.xml。

yarn-site.xml 用来配置 Yarn 的一些信息。yarn.nodemanager.aux-services 配置用户自定义服务，例如 MapReduce 的 shuffle。yarn.resourcemanager.address 设置客户端访问的地址，客户端通过该地址向 RM 提交应用程序、杀死应用程序等。yarn.resourcemanager.scheduler.adress 设置 ApplicationMaster 的访问地址，通过该地址向 ResourceManager 申请资源、释放资源等。yarn.resourcemanager.resource-tracker.address 设置 NodeManager 的访问地址，通过该地址向 ResourceManager 汇报心跳、领取任务等。yarn.resourcemanager.admin.address 设置管理员的访问地址，通过该地址向 ResourceManager 发送管理命令等。yarn.resourcemanager.webapp.address 设置对外 ResourceManager Web 访问地址，用户可通过该地址在浏览器中查看集群的各类信息。yarn-site.xml 配置信息如图 3-11 所示。

```
<configuration>
<!-- Site specific YARN configuration properties -->
    <property>
        <name>yarn.nodemanager.aux-services</name>
        <value>mapreduce_shuffle</value>
    </property>
    <property>
        <name>yarn.resourcemanager.address</name>
        <value>master:18040</value>
    </property>
    <property>
        <name>yarn.resourcemanager.scheduler.address</name>
        <value>master:18030</value>
    </property>
    <property>
        <name>yarn.resourcemanager.resource-tracker.address</name>
        <value>master:18025</value>
    </property>
    <property>
        <name>yarn.resourcemanager.admin.address</name>
        <value>master:18141</value>
    </property>
    <property>
        <name>yarn.resourcemanager.webapp.address</name>
        <value>master:18088</value>
    </property>
</configuration>
```

图 3-11　yarn-site.xml 配置

（6）创建文件 slaves。

　　touch /hadoop/hadoop-2.6.5/etc/hadoop/slaves

将其内容修改为各 slave 节点的主机名，如图 3-12 所示。

图 3-12　slaves 配置

通过该文件 master 节点知道集群中有几个子节点，然后通过主机名和/etc/hosts 的信息就可以知道各子节点对应的 IP，并和其通信。该文件只需在 master 节点配置即可，子节点不需要。

（7）将修改后的 Hadoop 分发到子节点。

```
scp -r /hadoop slave1:/
scp -r /hadoop slave2:/
```

（8）修改所有节点的环境变量。

在/etc/profile 文件末尾添加 Hadoop 的环境变量，在后续 HBase、ZooKeeper、Hive 等安装过程中仍需要执行类似操作。

```
export HADOOP_HOME=/hadoop/hadoop-2.6.5
export PATH=$PATH:$HADOOP_HOME/bin: $HADOOP_HOME/sbin
```

在配置环境变量的时候如果不小心配置错误，可能会造成无法正常启动到登录界面的情况。可以在系统刚启动后通过 Ctrl+Alt+F1 组合键切换到其他登录窗口，如 tty1，在里面输入管理员账号和密码，然后再对错误的环境配置进行修改即可。

（9）在 master 节点格式化 NameNode。

```
hdfs namenode -format
```

格式化后的显示内容如图 3-13 所示。

```
18/01/29 20:47:30 INFO namenode.FSNamesystem: HA Enabled: false
18/01/29 20:47:30 INFO namenode.FSNamesystem: Append Enabled: true
18/01/29 20:47:36 INFO util.GSet: Computing capacity for map INodeMap
18/01/29 20:47:36 INFO util.GSet: VM type       = 64-bit
18/01/29 20:47:36 INFO util.GSet: 1.0% max memory 966.7 MB = 9.7 MB
18/01/29 20:47:36 INFO util.GSet: capacity      = 2^20 = 1048576 entries
18/01/29 20:47:36 INFO namenode.NameNode: Caching file names occuring more than 10 times
18/01/29 20:47:36 INFO util.GSet: Computing capacity for map cachedBlocks
18/01/29 20:47:36 INFO util.GSet: VM type       = 64-bit
18/01/29 20:47:36 INFO util.GSet: 0.25% max memory 966.7 MB = 2.4 MB
18/01/29 20:47:36 INFO util.GSet: capacity      = 2^18 = 262144 entries
18/01/29 20:47:36 INFO namenode.FSNamesystem: dfs.namenode.safemode.threshold-pct = 0.9990000128746033
18/01/29 20:47:36 INFO namenode.FSNamesystem: dfs.namenode.safemode.min.datanodes = 0
18/01/29 20:47:36 INFO namenode.FSNamesystem: dfs.namenode.safemode.extension     = 30000
18/01/29 20:47:36 INFO namenode.FSNamesystem: Retry cache on namenode is enabled
18/01/29 20:47:36 INFO namenode.FSNamesystem: Retry cache will use 0.03 of total heap and retry cache entry expiry time is 600000 millis
18/01/29 20:47:36 INFO util.GSet: Computing capacity for map NameNodeRetryCache
18/01/29 20:47:36 INFO util.GSet: VM type       = 64-bit
18/01/29 20:47:36 INFO util.GSet: 0.029999999329447746% max memory 966.7 MB = 297.0 KB
18/01/29 20:47:36 INFO util.GSet: capacity      = 2^15 = 32768 entries
18/01/29 20:47:36 INFO namenode.NNConf: ACLs enabled? false
18/01/29 20:47:36 INFO namenode.NNConf: XAttrs enabled? true
18/01/29 20:47:36 INFO namenode.NNConf: Maximum size of an xattr: 16384
18/01/29 20:47:36 INFO namenode.FSImage: Allocated new BlockPoolId: BP-925896715-192.168.254.128-1517230056265
18/01/29 20:47:36 INFO common.Storage: Storage directory /hadoop/hdfs/tmp/dfs/name has been successfully formatted.
18/01/29 20:47:36 INFO namenode.FSImageFormatProtobuf: Saving image file /hadoop/hdfs/tmp/dfs/name/current/fsimage.ckpt_0000000000000000000 using no compression
18/01/29 20:47:36 INFO namenode.FSImageFormatProtobuf: Image file /hadoop/hdfs/tmp/dfs/name/current/fsimage.ckpt_0000000000000000000 of size 321 bytes saved in 0 seconds.
18/01/29 20:47:36 INFO namenode.NNStorageRetentionManager: Going to retain 1 images with txid >= 0
18/01/29 20:47:36 INFO util.ExitUtil: Exiting with status 0
18/01/29 20:47:36 INFO namenode.NameNode: SHUTDOWN_MSG:
/************************************************************
SHUTDOWN_MSG: Shutting down NameNode at master/192.168.254.128
```

图 3-13　NameNode 格式化信息

（10）master 节点启动 start-all.sh。

start-all.sh 也可由 start-dfs.sh 和 start-yarn.sh 代替。执行命令后，提示输入 yes/no 时，输入 yes。在 master 节点和 slave 节点分别输入 jps 命令，如果出现如图 3-14 至图 3-16 所示的进程表明 Hadoop 集群已经正常搭建。

```
[root@master hadoop]# jps         [root@slave1 hdfs]# jps        [root@slave2 hdfs]# jps
3040 ResourceManager              2157 DataNode                   2131 DataNode
2897 SecondaryNameNode            2254 NodeManager                2228 NodeManager
2709 NameNode                     2351 Jps                        2325 Jps
3358 Jps
```

图 3-14　master jps 进程　　　图 3-15　slave1 jps 进程　　　图 3-16　slave2 jps 进程

3.3　Hadoop 平台运行及验证

在配置 Hadoop 集群的时候，可能会不小心多次执行启动和格式化，这样会造成 NameNode 和 DataNode 中的 clusterID 值不一致，遇到这种情况修改 DataNode 中的 clusterID 即可，位置在/hadoop/hdfs/tmp/dfs/data/current/VERSION。

在初学的时候由于配置步骤不熟悉或者集群非正常关闭，会出现进程都已启动，但是 DataNode 为 0 的情况，可以通过 Web 访问查看集群情况，或者运行 Hadoop 自带的 MapReduce 实例验证集群节点情况，如果出现异常，可以根据每个节点下的 log 日志查看异常原因。通常可以通过以下方法解决：执行 stop-all.sh 命令，通过 jps 查看是否还有僵尸进程存在，有的话将其杀死；然后删除/hadoop/hdfs/tmp 目录下的所有文件，重新对其格式化。

在浏览器中输入地址 http://192.168.254.128:50070 检查 NameNode 和 DataNode 是否启动正常，如图 3-17 和图 3-18 所示。

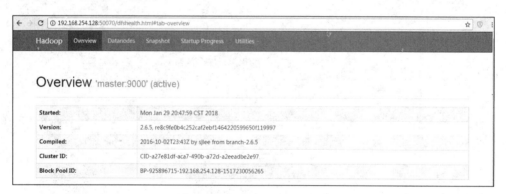

图 3-17　NameNode Web 信息

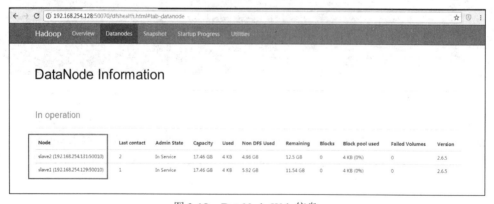

图 3-18　DataNode Web 信息

在浏览器中输入地址 http://192.168.254.128:18088 检查 Yarn 是否启动正常，如图 3-19 所示。

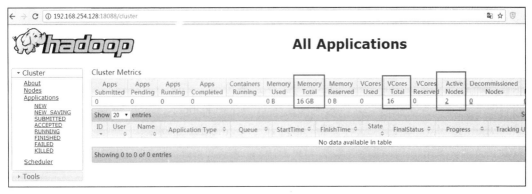

图 3-19　Yarn Web 信息

也可以在 master 节点上运行 Hadoop 自带的 MapReduce 例子，以验证整个集群是否正常。进入目录/hadoop/hadoop-2.6.5/share/hadoop/mapreduce/，运行命令：

  hadoop jar hadoop-mapreduce-examples-2.6.5.jar pi 10 10

上述代码是通过启动 10 个 map 和 10 个 reduce 任务求 pi 值，如果出现如图 3-20 所示的结果则表明整个集群正常。

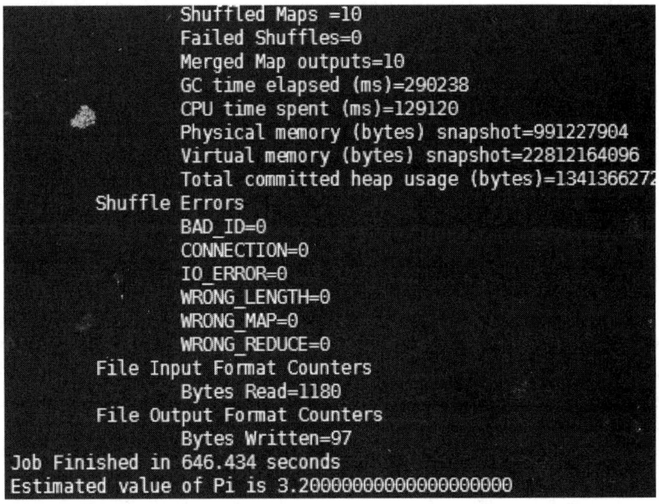

图 3-20　MapReduce pi 运行实例

3.4　本章小结

本章介绍的 Hadoop 搭建是分布式模式，通过该模式，可以快速地构建 Hadoop 的真实应用场景。Hadoop 是基于 Java 开发的，因此要搭建 JRE 运行环境，如果用户对 Linux 操作系统不熟悉的话，容易在网络配置、环境变量配置上出错，需要格外小心。Hadoop 集群搭建起来后，要进行验证，方能确保无误。

第 4 章 分布式文件系统 HDFS

在分布式文件系统中,一个文件会被保存到不同的计算机中,这就导致其读写机制和传统的本地文件系统大不相同。分布式文件系统 HDFS 是 Hadoop 平台的重要组成部分,本章将会从 HDFS 的架构、特点、工作机制和 shell 命令等几方面进行介绍。

4.1 HDFS 架构

分布式文件系统是指文件系统管理的物理存储资源不一定直接连接在本地节点上,而是通过计算机网络与节点相连。在分布式文件系统中,一个文件的数据虽然保存在不同的物理节点,但是对用户而言,其与普通的文件没有区别。

HDFS 作为一个可扩展的分布式文件系统,隐藏下层负载均衡、冗余复制等细节,对上层程序提供一个统一的文件系统 API 接口,具有存储管理、容错处理、高可扩展性、高可靠性和高可用性等特性。

4.1.1 HDFS 的基本框架

HDFS 主要由 3 个组件构成,分别是 Client、NameNode 和 DataNode。HDFS 是以主从模式运行的,其中 NameNode 运行在 master 节点,DataNode 运行在 slave 节点。HDFS 架构图如图 4-1 所示。

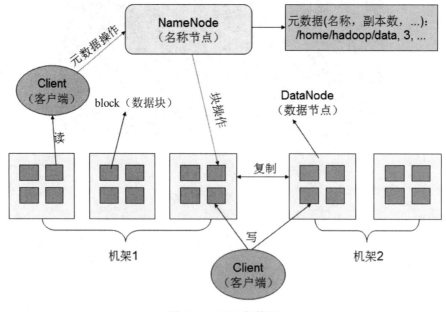

图 4-1 HDFS 架构图

1. Client（客户端）

用户可以通过类似 Linux shell 命令行方式或者 API 访问 HDFS，是用户操作数据的入口。Client 通过与 NameNode 交互，获取文件位置信息；与 DataNode 交互，实现读取和写入数据。

2. NameSpace（命名空间）

HDFS 采用传统的层次型文件组织结构，其结构和 Linux 文件系统很像，都有一个根节点"/"，所有文件都从该节点延伸而来，整个文件结构类似于树。

NameSpace 负责维护 HDFS 文件系统树及树内所有的文件和目录，这些信息以两种形式将文件永久保存在本地磁盘上：命名空间镜像（fsimage）和编辑日志（edits）。NameSpace 记录着每个数据块（包括备份数据块）位于哪个 DataNode 上，这些信息都保存在内存中，因此每次集群启动时会通过 DataNode 汇报重新生成。

NameSpace 由 NameNode 负责管理，所有对命名空间的改动（包括创建、删除、重命名、改变属性等，但是不包括打开、读取、写入数据）都会被 HDFS 记录下来并保存在 edits 中。

3. NameNode（名称节点）

在 Hadoop 1.x 版本中只有一个 NameNode，在 Hadoop 2.x 版本中可以设置多个 NameNode，但整个 Hadoop 集群只有一个 NameNode 处于 Active 状态。NameNode（通常简称为 NN）用来管理整个文件系统的命名空间和数据块映射信息，处理来自 Client 的文件访问请求（打开、关闭、重命名）和设置文件副本。

由于 NameNode 把文件系统的元数据信息保存在内存中，因此文件系统所能容纳的文件数目是由 NameNode 所在机器的内存大小决定的，这也决定了 Hadoop 集群的 DataNode 节点不能无限扩展。一般来说，一个文件、文件夹和数据块需要占据大约 150 字节左右的空间，假设集群有 100 万个文件，每一个文件占据一个数据块，就至少需要 300MB 内存，如果 NameNode 节点内存为 64GB，那么集群大概能存储两亿个文件信息。

4. DataNode（数据节点）

DataNode（通常简称为 DN）负责存储实际的数据，执行用户对数据的读写请求，响应 NameNode 对数据块的创建、删除指令，汇报数据块存储信息给 NameNode，以心跳包的形式向 NameSpace 发送存储的数据块列表。

DataNode 将 HDFS 数据以数据块的形式存储在本地文件系统的一个单独的文件中。实际上，DataNode 不会在同一个目录下创建所有的文件，它采用试探的方法来确定每个目录的最佳文件数目，并且在适当的时候创建子目录。

5. Secondary NameNode（第二名称节点）

帮助 NameNode 收集文件系统运行的状态信息，周期性（可以配置）保存 NameNode 的元数据（包括 fsimage 和 edits）。Secondary NameNode 是在文件系统中设置一个检查点来帮助 NameNode 更好地工作，但在 Hadoop 2.x 中采用了 HA 来取代它。

6. block（数据块）

为了提高系统的读写效率，HDFS 中采用了数据块作为数据读写单位。一个文件可以分成若干块，每个块可以保存在不同的 DataNode 上，这样 HDFS 就可以存储比本地磁盘容量大得多的文件，数据块对用户是透明的，用户看到的仍是一个完整文件。

与一般文件系统中大小为几 KB 的数据块不同，HDFS 数据块的默认大小是 128MB（在 Hadoop 1.0 中是 64MB），根据实际生产的需要，还可以设置成 256MB 甚至更大。HDFS 之所以设置这么大，目的是为了减少寻址开销。由于 HDFS 存储大文件较多，如果数据块设置太小，就需要读取大量数据块，而数据块又是分布式存储的，会造成硬盘寻址时间比传输时间还长。为了减少寻址时间提高系统吞吐量，数据块不能设置太小。数据表明，如果寻址时间在 10 毫秒左右，传输速率是 100Mb/s，为了使寻址时间为传输时间的 1%，则需要 100MB 左右的块大小。数据块大的另一个好处就是一个大文件会由较少的数据块组成，减少 NameNode 的内存消耗，增加集群存储文件的数目。

需要注意的是，数据块只是 HDFS 中文件分割的单位，并不是数据存储单位。比如一个文件为 200MB，它会被分成两个数据块，一个数据块为 128MB，另一个数据块为 72MB 而不是 128MB。

4.1.2 HDFS 的特点

HDFS 是 Hadoop 项目的核心子项目，是分布式计算中数据存储管理的基础，是基于流数据模式访问和处理超大文件的需求而开发的，可以运行于廉价的商用服务器上。

1. 优点

（1）高容错性。

1）上传的数据自动保存多个副本。它是通过增加副本的数量来增加它的容错性的。

2）如果某一个副本丢失，HDFS 机制会复制其他机器上的副本，而我们不必关注它的实现。

（2）适合大数据的处理。

1）能够处理 GB、TB，甚至 PB 级别的数据。

2）能够处理百万规模的数据，数量非常的大。

（3）流式文件写入。

1）一次写入，多次读取。文件一旦写入，不能修改，只能增加。

2）这样可以保证数据的一致性。

（4）可构建在廉价机器上。

1）通过多副本提高可靠性。

2）提供了容错和恢复机制。

2. 劣势

（1）低延时数据访问。

它适合高吞吐率的场景，就是在某一时间内写入大量的数据。但是它在低延时的情况下是不行的，比如毫秒级以内读取数据，这样它是很难做到的。

（2）小文件的存储。

1）存放大量小文件的话，它会占用 NameNode 的大量内存在存储文件、目录和块信息上。这样是不可取的，因为 NameNode 的内存总是有限的。

2）小文件存储的寻道时间会超过文件的读取时间，这违背了 HDFS 的设计初衷。

（3）并发写入、文件随机修改。

1）一个文件只能一个线程写，不能多个线程同时写。

2）仅支持文件的追加，不支持文件的随机修改。

（4）不支持超强的事务。

没有像关系型数据库那样，对事务具有强有力的支持。

4.2 HDFS 的工作机制

HDFS 集群分为两大角色：NameNode 和 DataNode。NameNode 负责管理整个文件系统的元数据，DataNode 负责管理用户的文件数据块。文件会按照固定的大小（blocksize）切成若干块后分布式存储在若干台 DataNode 上，每一个数据块可以有多个副本，并存放在不同的 DataNode 上。DataNode 会定期向 NameNode 汇报自身所保存的文件 block 信息，而 NameNode 则会负责保持文件的副本数量。HDFS 的内部工作机制对客户端保持透明，客户端请求访问 HDFS 都是通过向 NameNode 申请来进行。

4.2.1 HDFS 读写过程分析

1. HDFS 写数据流程

客户端要向 HDFS 写数据，首先要跟 NameNode 通信以确认可以写文件并获得接收文件 block 的 DataNode，然后客户端按顺序将文件逐个 block 传递给相应 DataNode，并由接收到 block 的 DataNode 负责向其他 DataNode 复制 block 的副本。详细步骤如下：

（1）客户端向 NameNode 发送上传文件请求，NameNode 对要上传的目录和文件进行检查，判断是否可以上传，并向客户端返回检查结果。

（2）客户端得到上传文件的允许后读取客户端配置，如果没有指定配置则会读取默认配置（例如副本数和块大小默认为 3 和 128MB，副本是由客户端决定的），向 NameNode 请求上传一个数据块。

（3）NameNode 会根据客户端的配置来查询 DataNode 信息，如果使用默认配置，那么最终结果会返回同一个机架的两个 DataNode 和另一个机架的 DataNode，这称为"机架感知"策略。

机架感知：HDFS 采用一种称为机架感知（rack-aware）的策略来改进数据的可靠性、可用性和网络带宽的利用率。大型 HDFS 实例一般运行在跨越多个机架的计算机组成的集群上，不同机架上的两台机器之间的通信需要经过交换机。在大多数情况下，同一个机架内的两台机器间的带宽会比不同机架的两台机器间的带宽大。通过一个机架感知的过程，NameNode 可以确定每个 DataNode 所属的机架 id。一个简单但没有优化的策略就是将副本存放在不同的机架上，这样可以有效防止当整个机架失效时数据的丢失，并且允许读数据的时候充分利用多个机架的带宽。这种策略设置可以将副本均匀分布在集群中，有利于当组件失效情况下的负载均衡。但是，因为这种策略的一个写操作需要传输数据块到多个机架，这增加了写的代价。在大多数情况下，副本系数是 3，HDFS 的存放策略是将一个副本存放在本地机架的节点上，一个副本放在同一机架的另一个节点上，最后一个副本放在不同机架的节点上。这种策略减少了机架间的数据传输，这就提高了写操作的效率。机架的错误远远比节点的错误少，所以这个策略不会影响到数据的可靠性和可用性。与此同时，因为数据块只放在两个（不是三个）不同的机架上，所以此策略减少了读取数据时需要的网络传输总带宽。在这种策略下，副本并不是均匀分布在不同的机架上。三分之一的副本在一个节点上，三分之二的副本在一个机架上，其他副本均匀分布在剩下的机架中，这一策略在不损害数据可靠性和读取性能的情况下改进了写的性能。

（4）客户端在开始传输数据块之前会把数据缓存在本地，当缓存大小超过了一个数据块的大小后，客户端就会从 NameNode 获取要上传的 DataNode 列表。之后会在客户端和第一个 DataNode 建立连接开始流式地传输数据，这个 DataNode 会一小部分一小部分（4KB）地接收数据然后写入本地仓库，同时会把这些数据传输到第二个 DataNode，第二个 DataNode 也同样一小部分一小部分地接收数据并写入本地仓库，同时传输给第三个 DataNode，依此类推。这样逐级调用和返回之后，待这个数据块传输完成客户端会告诉 NameNode 数据块传输完成，这时候 NameNode 才会更新元数据信息记录操作日志。

（5）第一个数据块传输完成后会使用同样的方式传输下面的数据块直到整个文件上传完成。

1）请求和应答是使用 RPC 的方式，客户端通过 Client Protocol 与 NameNode 通信，NameNode 和 DataNode 之间使用 DataNode Protocol 交互。在设计上，NameNode 不会主动发起 RPC，而是响应来自客户端或 DataNode 的 RPC 请求。客户端和 DataNode 之间是使用 Socket 进行数据传输，与 NameNode 之间的交互采用 nio 封装的 RPC。

2）HDFS 有自己的序列化协议。

3）在数据块传输成功后但客户端没有告诉 NameNode 之前，如果 NameNode 死机，那么这个数据块就会丢失。

4）在流式复制时，逐级传输和响应采用响应队列来等待传输结果，队列响应完成后返回给客户端。

5）在流式复制时如果有一台或两台（不是全部）没有复制成功，不影响最后的结果，只不过 DataNode 会定期向 NameNode 汇报自身信息。如果发现异常 NameNode 会指挥 DataNode 删除残余数据和完善副本，如果副本数量少于某个最小值就会进入安全模式。

安全模式：NameNode 启动后会进入一个称为安全模式的特殊状态，处于安全模式的 NameNode 是不会进行数据块复制的。NameNode 从所有的 DataNode 接收心跳信号和块状态报告，块状态报告包括了某个 DataNode 所有的数据块列表。每个数据块都有一个指定的最小副本数，当 NameNode 检测确认某个数据块的副本数目达到这个最小值时该数据块就会被认为是副本安全（safely replicated）的；在一定百分比（这个参数可以配置）的数据块被 NameNode 检测确认是安全之后（加上一个额外的 30 秒等待时间），NameNode 将退出安全模式状态。接下来它会确定还有哪些数据块的副本没有达到指定数目，并将这些数据块复制到其他 DataNode 上。

2. HDFS 读数据流程

客户端将要读取的文件路径发送给 NameNode，NameNode 获取文件的元信息（主要是 block 的存放位置信息）返回给客户端，客户端根据返回的信息找到相应 DataNode 逐个获取文件的 block 并在客户端本地进行数据追加合并从而获得整个文件。详细步骤如下：

（1）客户端向 NameNode 发起 RPC 调用，请求读取文件数据。

（2）NameNode 检查文件是否存在，如果存在则获取文件的元信息（block id 以及对应的 DataNode 列表）。

（3）客户端收到元信息后选取一个网络距离最近的 DataNode，依次请求读取每个数据块。客户端首先要校验文件是否损坏，如果损坏，客户端会选取另外的 DataNode 请求。

（4）DataNode 与客户端建立 Socket 连接，传输对应的数据块，客户端收到数据缓存到本地，之后写入文件。

（5）依次传输剩下的数据块，直到整个文件合并完成。

从某个 DataNode 获取的数据块有可能是损坏的，损坏可能是由 DataNode 的存储设备错误、网络错误或者软件 bug 造成的。HDFS 客户端软件实现了对 HDFS 文件内容的校验和（checksum）检查。当客户端创建一个新的 HDFS 文件时，将会计算这个文件每个数据块的校验和，并将校验和作为一个单独的隐藏文件保存在同一个 HDFS 名字空间下。当客户端获取文件内容后，它会检验从 DataNode 获取的数据跟相应的校验和文件中的校验和是否匹配，如果不匹配，客户端可以选择从其他 DataNode 获取该数据块的副本。

3. HDFS 删除数据

HDFS 删除数据的流程相对简单，详细步骤如下：

（1）客户端向 NameNode 发起 RPC 调用，请求删除文件，NameNode 检查合法性。

（2）NameNode 查询文件相关元信息，向存储文件数据块的 DataNode 发出删除请求。

（3）DataNode 删除相关数据块，返回结果。

（4）NameNode 返回结果给客户端。

当用户或应用程序删除某个文件时，这个文件并没有立刻从 HDFS 中删除。实际上，HDFS 会将这个文件重命名后转移到/trash 目录中。只要文件还在/trash 目录中，该文件就可以被迅速地恢复，/trash 目录仅仅保存被删除文件的最后副本，/trash 目录与其他的目录没有什么区别。文件在/trash 目录中保存的时间是可配置的，目前的默认策略是删除/trash 目录中保留时间超过 6 小时的文件，当超过这个时间后，NameNode 就会将该文件从名字空间中删除，删除文件会使得该文件相关的数据块被释放。

当一个文件的副本系数被减小后，NameNode 会将过剩的副本删除。下次心跳检测时会将该信息传递给 DataNode。DataNode 随即移除相应的数据块，集群中的空闲空间加大。同样，在调用 setReplication API 结束和集群中空闲空间增加会有一定的延迟。

4.2.2 NameNode 的工作机制

NameNode 主要负责客户端请求的响应和元数据的管理（查询、修改）。NameNode 对数据的管理采用了三种存储形式：内存元数据（NameSystem）、磁盘元数据镜像文件、数据操作日志文件（可通过日志运算出元数据）。

1. NameNode 元数据的目录结构

NameNode 元数据相关的目录结构配置在 hdfs-site.xml 上的 dfs.name.dir 项，具体目录为 $dfs.name.dir/current。在该目录下主要有 VERSION、edits、fsimage 和 seen_txid 等文件，其目录结构如图 4-2 所示。

（1）VERSION。

VERSION 文件是 Java 属性文件，内容如下：

```
#Mon Jan 29 20:47:36 CST 2018
namespaceID=1090922262
clusterID=CID-a27e81df-aca7-490b-a72d-a2eeadbe2e97
cTime=0
storageType=NAME_NODE
blockpoolID=BP-925896715-192.168.254.128-1517230056265
layoutVersion=-60
```

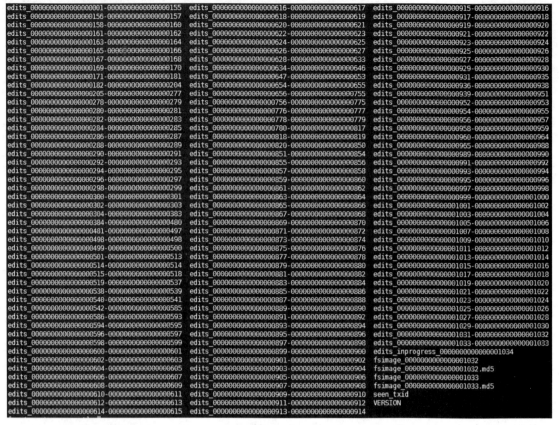

图 4-2　NameNode 元数据目录结构

1）namespaceID：是文件系统的唯一标识符，在文件系统首次格式化之后生成。

2）cTime：表示 NameNode 存储时间的创建时间，由于 NameNode 没有更新过，所以这里的记录值为 0，以后对 NameNode 升级之后，cTime 将会记录更新时间戳。

3）storageType：说明这个文件存储的是什么类型的数据结构信息（如果是 DataNode，storageType=DATA_NODE）。

4）blockpoolID：是针对每一个 NameSpace 所对应的 blockpool 的 ID，上面的这个=BP-925896715-192.168.254.128-1517230056265 就是在 master 的 NameSpace 下的存储块池的 ID，这个 ID 包括了其对应的 NameNode 节点的 IP 地址。

5）layoutVersion：表示 HDFS 永久性数据结构的版本信息，只要数据结构变更，版本号也要递减，此时的 HDFS 也需要升级，否则磁盘仍旧是使用旧版本的数据结构，这会导致新版本的 NameNode 无法使用。

6）clusterID：是系统生成或者手动指定的集群 ID，在-clusterid 选项中可以使用它：

①$HADOOP_HOME/bin/hdfs namenode -format [-clusterId <cluster_id>]

选择一个唯一的 cluster_id，并且这个 cluster_id 不能与环境中的其他集群有冲突。如果没有提供 cluster_id，则会自动生成一个唯一的 ClusterID。

②使用如下命令格式化其他 Namenode：

　　$HADOOP_HOME/bin/hdfs namenode -format -clusterId <cluster_id>

③升级集群至最新版本。

在升级过程中需要提供一个 ClusterID，例如：

    $HADOOP_PREFIX_HOME/bin/hdfs start namenode --config
    $HADOOP_CONF_DIR -upgrade -clusterId <cluster_ID>

如果没有提供 ClusterID，则会自动生成一个 ClusterID。

（2）fsimage 和 edits。

fsimage 和 edits 是 NameNode 中两个很重要的文件。fsimage 镜像文件包含了整个 HDFS 文件系统的所有目录、文件的 indoe 信息和数据块到文件的映射。对于目录来说包括修改时间、访问权限控制信息（目录所属用户、所在组等）；对于文件来说包括数据块描述信息、修改时间、访问时间等。

edits 文件主要是在 NameNode 已经启动的情况下对 HDFS 进行的各种更新操作进行记录，HDFS 客户端执行所有的写操作都会被记录到 edits 文件中。例如，在 HDFS 中创建一个文件，NameNode 就会在 edits 中插入一条记录来表示，同样地，修改文件的副本系数也将往 edits 插入一条记录。

NameNode 元数据是保存在内存中的，由 DataNode 向其汇报每个节点所保存的相关信息，当用户访问文件时，通过 NameNode 就可以知道到哪个 DataNode 访问哪些数据块。NameNode 元数据的逻辑关系如图 4-3 所示。

```
file1.txt=(Blk1:DN1, DN2, DN3),
         (Blk2:DN2, DN3, DN4),

file2.txt=(Blk1:DN1, DN2, DN4),
         (Blk2:DN2, DN3, DN5),
         (Blk3:DN1, DN3, DN4)
```

图 4-3　元数据逻辑关系示意图

file1.txt 文件占用两个块，分别是 BlkA 和 BlkB。BlkA 的三个副本分别保存在 DN1、DN2、DN3 上，BlkB 的三个副本分别保存在 DN2、DN3、DN4 上。至于读取 file1.txt 的时候会分别到哪个 DataNode 上读取，4.2.1 节会有详细介绍，这里不再赘述。

（3）seen_teid 文件。

存放 transactionid 的文件，format 之后是 0，代表的是 NameNode 里面的 edits_*文件的尾数，NameNode 重启的时候，会按照 seen_txid 的数字，顺序从头跑 edits_0000001 到 seen_txid 的数字。所以当 HDFS 发生异常重启的时候，一定要比对 seen_txid 内的数字是不是 edits 最后的尾数，不然会发生构造 NameNode 时 metaData 的资料缺少，导致误删 DataNode 上多余的 block。

文件中记录的是 edits 滚动的序号，每次重启 NameNode 时，NameNode 就知道要将哪些 edits 进行加载 edits。

（4）current 目录。

current 目录下在 format 的同时也会生成 fsimage 和 edits 文件及其对应的 md5 校验文件。

2. 元数据存储机制

当 NameNode 启动时，从本地文件系统中读取 edits 和 fsimage 文件，将所有 edits 中的事务作用在内存中的 fsimage 上，并将这个新版本的 fsimage 从内存中写入到本地文件上，然后删除旧的 edits（这个过程称为检查点），最后等待各个 DataNode 向 NameNode 汇报数据块的信息来组装 block id 映射关系。

DataNode 启动时会扫描本地文件系统，产生一个本地文件对应的 HDFS 数据块的列表（每个数据块会对应一个本地文件），然后作为报告发送到 NameNode（这个报告称为块状态报告）。NameNode 在接收到每个 DataNode 的块汇报信息后，将其同所在的 DataNode 信息等保存在内存中。

在 Hadoop 1.x 中，如果 NameNode 失效，可以通过 Secondary NameNode 中保存的 fsimage 和 edits 数据恢复出 NameNode 最近的状态。为了加快 NameNode 重启速度，Secondary NameNode 还会定期合并 edits。

在 Hadoop 2.x 中不再使用 Secondary NameNode 作为 NameNode 的恢复手段，而是采用了 HA 机制。在 Hadoop 集群中有多台 NameNode，但只有一台 NameNode 处于 Active 状态，其余的均处于 Standby 状态。Active NameNode 负责集群中所有客户端的操作，而 Standby NameNode 主要用于备用，它主要维持足够的状态，如果必要，还可以提供快速的故障恢复。

为了让 Standby NameNode 的状态和 Active NameNode 保持同步，即元数据保持一致，会通过一组称为 JournalNode（JournalNode 至少部署在三个节点上，而且必须是奇数）的独立进程进行相互通信。当 Active NameNode 的命名空间有任何修改时，它需要持久化到一半以上的 JournalNode 上（通过 edits 持久化存储）。Standby NameNode 有能力读取 JournalNode 中的变更信息，并且一直监控 edits 的变化，并把变化应用于自己的命名空间。Standby NameNode 可以确保在集群出错时命名空间状态已经完全同步，然后就可以切换到 Active 状态。

4.2.3 元数据的 CheckPoint

每隔一段时间，会由 Secondary NameNode 将 NameNode 上积累的所有 edits 和一个最新的 fsimage 下载到本地，并加载到内存进行 merge（这个过程称为 CheckPoint）。

1. CheckPoint 的详细过程

（1）从 Secondary NameNode 节点通知 NameNode 进行 CheckPoint。

（2）NameNode 切换出新的日志文件，以后的日志都写到新的日志文件中。

（3）从 Secondary NameNode 节点下载 fsimage 文件及旧的日志文件，fsimage 文件只有第一次下载，以后只需要传输 edits 日志文件。

（4）Secondary NameNode 节点将 fsimage 文件加载到内存中，并将日志文件与 fsimage 合并，然后生成新的 fsimage 文件。

（5）从 Secondary NameNode 节点将新的 fsimage 文件传回 NameNode 节点。

（6）NameNode 节点可以将旧的 fsimage 文件及旧的日志文件切换为新的 fsimage 和 edits 日志文件并更新 fstime 文件，写入此次 CheckPoint 的时间。

这样，NameNode 中只需要花费很少的时间，始终保持了最新的元数据信息，由于有了备份机制，即使死机并重启后还是可以恢复元数据，但最新的日志可能来不及同步会有丢失。

2. CheckPoint 配置参数

（1）hdfs-site.xml 文件。

  dfs.namenode.checkpoint.check.period=60 #检查触发条件是否满足的频率，60 秒
  dfs.namenode.checkpoint.dir=file://${hadoop.tmp.dir}/dfs/namesecondary

以上两个参数作 CheckPoint 操作时是 Secondary NameNode 的本地工作目录。

  dfs.namenode.checkpoint.edits.dir=${dfs.namenode.checkpoint.dir}
  dfs.namenode.checkpoint.max-retries=3 #最大重试次数
  dfs.namenode.checkpoint.period=3600 #两次 CheckPoint 之间的时间间隔，3600 秒
  dfs.namenode.checkpoint.txns=1000000 #两次 CheckPoint 之间最大的操作记录

（2）CheckPoint 的附带作用。

NameNode 和 Secondary NameNode 的工作目录存储结构完全相同，所以，当 NameNode 故障退出需要重新恢复时，可以从 Secondary NameNode 的工作目录中将 fsimage 拷贝到 NameNode 的工作目录，以恢复 NameNode 的元数据。

4.2.4 DataNode 的工作机制

DataNode 的工作职责：存储管理用户的数据块数据，定期向 NameNode 汇报自身所持有的 block 信息（通过心跳信息上报）。

DataNode 进程死亡或者网络故障造成 DataNode 无法与 NameNode 通信，NameNode 不会立即把该节点判定为死亡，要经过一段时间，这段时间暂称为超时时长。HDFS 默认的超时时长为 10 分钟+30 秒。配置超时如下：

```
<property>
    <name>dfs.blockreport.intervalMsec</name>
    <value>3600000</value>
    <description>Determines block reporting interval in milliseconds.</description>
</property>
```

如果定义超时时长为 timeout，则超时时长的计算公式为：

timeout = 2 * heartbeat.recheck.interval + 10 * dfs.heartbeat.interval

而默认的 heartbeat.recheck.interval 大小为 5 分钟，dfs.heartbeat.interval 默认为 3 秒。需要注意的是，hdfs-site.xml 配置文件中的 heartbeat.recheck.interval 的单位为毫秒，dfs.heartbeat.interval 的单位为秒。举个例子，如果 heartbeat.recheck.interval 设置为 5000（毫秒），dfs.heartbeat.interval 设置为 3（秒，默认），则总的超时时长为 40 秒。

```
<property>
    <name>heartbeat.recheck.interval</name>
    <value>2000</value>
</property>

<property>
    <name>dfs.heartbeat.interval</name>
    <value>1</value>
</property>
```

4.3 HDFS shell 命令

HDFS 的访问方式有两种：一种是 HDFS shell，一种是 Java API 方式。HDFS shell 命令应使用 hadoop fs 或 hdfs dfs（其中官网建议使用 hdfs dfs 方式访问）。所有的 HDFS shell 命令使用 URI 路径作为参数，URI 格式是 scheme://path，对于 HDFS 文件系统，scheme 是 hdfs；对于本地文件系统，scheme 是 file。其中 scheme 是可选的，如果未加指定，默认是 HDFS 文件系统。一个 HDFS 文件或目录如/parent/path 可以表示成 hdfs://parent/path，或者是更简单的/parent/path。

启动 HDFS 后，输入 hdfs fs 命令，即可显示 HDFS 常用命令的用法。

```
[root@master /]# hdfs dfs
Usage: hadoop fs [generic options]
        [-appendToFile <localsrc> ... <dst>]
        [-cat [-ignoreCrc] <src> ...]
        [-checksum <src> ...]
        [-chgrp [-R] GROUP PATH...]
        [-chmod [-R] <MODE[,MODE]... | OCTALMODE> PATH...]
        [-chown [-R] [OWNER][:[GROUP]] PATH...]
        [-copyFromLocal [-f] [-p] <localsrc> ... <dst>]
        [-copyToLocal [-p] [-ignoreCrc] [-crc] <src> ... <localdst>]
        [-count [-q] <path> ...]
        [-cp [-f] [-p | -p[topax]] <src> ... <dst>]
        [-createSnapshot <snapshotDir> [<snapshotName>]]
        [-deleteSnapshot <snapshotDir> <snapshotName>]
        [-df [-h] [<path> ...]]
        [-du [-s] [-h] <path> ...]
        [-expunge]
        [-get [-p] [-ignoreCrc] [-crc] <src> ... <localdst>]
        [-getfacl [-R] <path>]
        [-getfattr [-R] {-n name | -d} [-e en] <path>]
        [-getmerge [-nl] <src> <localdst>]
        [-help [cmd ...]]
        [-ls [-d] [-h] [-R] [<path> ...]]
        [-mkdir [-p] <path> ...]
        [-moveFromLocal <localsrc> ... <dst>]
        [-moveToLocal <src> <localdst>]
        [-mv <src> ... <dst>]
        [-put [-f] [-p] <localsrc> ... <dst>]
        [-renameSnapshot <snapshotDir> <oldName> <newName>]
        [-rm [-f] [-r|-R] [-skipTrash] <src> ...]
        [-rmdir [--ignore-fail-on-non-empty] <dir> ...]
        [-setfacl [-R] [{-b|-k} {-m|-x <acl_spec>} <path>]|[--set <acl_spec> <path>]]
        [-setfattr {-n name [-v value] | -x name} <path>]
        [-setrep [-R] [-w] <rep> <path> ...]
```

[-stat [format] <path> ...]
[-tail [-f] <file>]
[-test -[defsz] <path>]
[-text [-ignoreCrc] <src> ...]
[-touchz <path> ...]
[-usage [cmd ...]]

Generic options supported are
-conf <configuration file>   specify an application configuration file
-D <property=value>   use value for given property
-fs <local|namenode:port>   specify a namenode
-jt <local|jobtracker:port>   specify a job tracker
-files <comma separated list of files>   specify comma separated files to be copied to the map reduce cluster
-libjars <comma separated list of jars>   specify comma separated jar files to include in the classpath.
-archives <comma separated list of archives>   specify comma separated archives to be unarchived on the compute machines.

The general command line syntax is
bin/hadoop command [genericOptions] [commandOptions]

4.3.1 帮助相关命令

1. usage
查看命令的用法，例如查看 ls 的用法：
```
[root@master /]# hdfs dfs -usage ls
Usage: hadoop fs [generic options] -ls [-d] [-h] [-R] [<path> ...]
```

2. help
查看命令的详细帮助，例如查看 ls 命令的帮助：
```
[root@master /]# hdfs dfs -help ls
-ls [-d] [-h] [-R] [<path> ...] :
  List the contents that match the specified file pattern. If path is not
  specified, the contents of /user/<currentUser> will be listed. Directory entries
  are of the form:
      permissions - userId groupId sizeOfDirectory(in bytes)
  modificationDate(yyyy-MM-dd HH:mm) directoryName

  and file entries are of the form:
      permissions numberOfReplicas userId groupId sizeOfFile(in bytes)
  modificationDate(yyyy-MM-dd HH:mm) fileName

  -d  Directories are listed as plain files.
  -h  Formats the sizes of files in a human-readable fashion rather than a number
      of bytes.
  -R  Recursively list the contents of directories.
```

4.3.2 查看相关命令

1. ls

查看文件或目录。下例中，hdfs://master:9000 是 fs.defaultFS 配置的值，hdfs://master:9000/ 即表示 HDFS 文件系统中的根目录，如果使用的是 HDFS 文件系统，可以简写为/。

```
[root@master /]# hdfs dfs -ls   hdfs://master:9000/
Found 4 items
drwxr-xr-x   - root supergroup          0 2017-08-31 10:16 hdfs://master:9000/hbase
drwxrwxrwx   - root supergroup          0 2018-01-13 16:36 hdfs://master:9000/test
drwxrwxrwx   - root supergroup          0 2017-08-24 17:28 hdfs://master:9000/tmp
drwxr-xr-x   - root supergroup          0 2017-08-26 11:19 hdfs://master:9000/user
[root@master /]# hdfs dfs -ls /
Found 4 items
drwxr-xr-x   - root supergroup          0 2017-08-31 10:16 /hbase
drwxrwxrwx   - root supergroup          0 2018-01-13 16:36 /test
drwxrwxrwx   - root supergroup          0 2017-08-24 17:28 /tmp
drwxr-xr-x   - root supergroup          0 2017-08-26 11:19 /user
```

选项-R：连同子目录的文件一起列出，例如：

```
[root@master /]# hdfs dfs -ls -R /user
drwxr-xr-x   - root supergroup          0 2017-08-26 11:19 /user/hive
drwxr-xr-x   - root supergroup          0 2017-08-26 11:19 /user/hive/warehouse
drwxr-xr-x   - root supergroup          0 2017-08-26 11:29 /user/hive/warehouse/t1
-rw-r--r--   2 root supergroup         27 2017-08-26 11:29 /user/hive/warehouse/t1/t1.txt
drwxr-xr-x   - root supergroup          0 2017-08-29 17:48 /user/root
drwxr-xr-x   - root supergroup          0 2017-08-29 18:05 /user/root/.sparkStaging
```

2. cat

显示文件内容，例如查看/user/hive/warehouse/t1/中 t1.txt 的文件内容：

```
[root@master /]# hdfs dfs -cat /user/hive/warehouse/t1/t1.txt
1    2    3
4    5    6
7    8    9
10   11   12
```

3. text

将给定的文件以文本的格式输出，允许的格式有 zip、TextRecordInputStream、Avro。当文件为文本文件时，等同于 cat。例如查看/user/hive/warehouse/t1/中 t1.txt 的文件内容：

```
[root@master /]# hdfs dfs -text /user/hive/warehouse/t1/t1.txt
1    2    3
4    5    6
7    8    9
10   11   12
```

4. tail

显示文件最后 1KB 的内容。选项-f 为当文件内容增加时显示追加的内容。例如查看文件末尾内容如图 4-4 所示。

图 4-4　d0bc845af01c4ef7a648018e22ab072f 文件末尾内容

5. checksum

显示文件的校验和信息。因为需要和存储文件每个块的 DataNode 相互通信，因此对大量的文件使用此命令效率可能会低。例如查看 /user/hive/warehouse/t1/t1.txt 的 MD5 校验和信息：

```
[root@master /]# hdfs dfs -checksum /user/hive/warehouse/t1/t1.txt
/user/hive/warehouse/t1/t1.txt    MD5-of-0MD5-of-512CRC32C
000002000000000000000000a09cda2a56f5af74e88d167a7e66435f
```

4.3.3　文件及目录相关命令

1. touchz

创建一个空文件，如果存在指定名称的非空文件，则返回错误。

```
[root@master /]# hdfs dfs -touchz /user/hive/warehouse/t1/t1.txt
touchz: `/user/hive/warehouse/t1/t1.txt': Not a zero-length file      --非空时给出错误提示
[root@master /]# hdfs dfs -touchz /1.txt
[root@master /]# hdfs dfs -ls /
Found 5 items
-rw-r--r--    2 root supergroup      0 2018-01-13 17:12 /1.txt    --创建成功
drwxr-xr-x    - root supergroup      0 2017-08-31 10:16 /hbase
drwxrwxrwx    - root supergroup      0 2018-01-13 16:36 /test
drwxrwxrwx    - root supergroup      0 2017-08-24 17:28 /tmp
```

2. appendToFile

向现有文件中追加内容，例如：

```
[root@master /]# hdfs dfs -text /1.txt
[root@master /]#
[root@master /]# hdfs dfs -appendToFile /user/hive/warehouse/t1/t1.txt /1.txt
```

appendToFile: /user/hive/warehouse/t1/t1.txt（没有那个文件或目录）

通过之前的命令知道该文件是存在的，可为什么有这个提示呢，原来 HDFS 设计之初并不支持给文件追加内容，将 hdfs-site.xml 中的以下属性修改为 true 即可：

```
<property>
    <name>dfs.support.append</name>
    <value>true</value>
</property>
```

将 HDFS 服务重启后再次执行。

```
[root@master /]# hdfs dfs -text /1.txt
1    2    3
4    5    6
7    8    9
10   11   12
```

可见能够添加成功。

3. put

从本地文件系统上传文件到 HDFS，例如：

```
[root@master /]# hdfs dfs -put /usr/log.txt /
[root@master /]# hdfs dfs -cat /log.txt       --查看上传后的文件内容
Hadoop
```

选项-f 为如果文件已经存在，则覆盖已有文件。

```
[root@master /]# hdfs dfs -put /usr/log.txt /
put: `/log.txt': File exists      --文件已存在时给出错误提示
[root@master /]# hdfs dfs -put -f /usr/log.txt /
[root@master /]#           --使用-f 选项后没有再报错
```

选项-p 为保留原文件的访问和修改时间、用户和组、权限属性。

```
[root@master /]# chmod 777 /usr/log.txt     --修改权限为 rwxrwxrwx
[root@master /]# hdfs dfs -put /usr/log.txt /
[root@master /]# hdfs dfs -ls /log.txt
-rw-r--r--    2 root supergroup      0 2018-01-13 17:26 /log.txt
```

不使用-p 选项，上传后文件属性。

```
[root@master /]# hdfs dfs -put -f -p /usr/log.txt /
[root@master /]# hdfs dfs -ls /log.txt
-rwxrwxrwx    2 root root            0 2018-01-13 17:26 /log.txt
```

使用-p 选项，上传后文件属性。

4. get

从 HDFS 上下载文件到本地，与 put 不同，没有覆盖本地已有文件的选项。

```
[root@master /]# hdfs dfs -get /log.txt ~
[root@master /]$ cat ~/input1.txt   --查看本地下载的文件
Hadoop
```

5. getmerge

将指定的 HDFS 中源目录下的文件合并成一个文件并下载到本地，源文件保留。

```
[root@master /]# hdfs dfs -text /input/input1.txt
Hadoop     --input1.txt 内容
[root@master /]# hdfs dfs -text /input/input2.txt
```

```
hello Hadoop       --input2.txt 内容
[root@master /]# hdfs dfs -getmerge /input/ ~/merge.txt
[root@master /]$ cat ~/merge.txt
Hadoop
hello Hadoop       --合并后本地文件的内容
```

选项-nl 为在每个文件的最后增加一个新行。

```
[root@master /]# hdfs dfs -getmerge -nl /input/ ~/merge.txt
[root@master /]# cat ~/merge.txt
Hadoop
        --input1.txt 增加的新行
hello Hadoop
        --input2.txt 增加的新行
```

6. copyFromLocal

从本地文件系统上传文件到 HDFS，与 put 命令相同。

7. copyToLocal

从 HDFS 下载文件到本地文件系统，与 get 命令相同。

8. moveFromLocal

与 put 命令相同，只是上传成功后本地文件会被删除。

9. mv

同 Linux 的 mv 命令，移动或重命名文件，例如：

```
[root@master /]# hdfs dfs -ls /
[root@master /]# hdfs dfs -mv /1.txt /user    --移动文件
[root@master /]# hdfs dfs -ls /
Found 5 items
drwxr-xr-x   - root supergroup          0 2017-08-31 10:16 /hbase
-rw-r--r--   2 root root                0 2018-01-13 17:26 /log.txt
drwxrwxrwx   - root supergroup          0 2018-01-13 16:36 /test
drwxrwxrwx   - root supergroup          0 2017-08-24 17:28 /tmp
drwxr-xr-x   - root supergroup          0 2018-01-13 17:34 /user
[root@master /]# hdfs dfs -ls /user
Found 3 items
-rw-r--r--   2 root supergroup          0 2018-01-13 17:16 /user/1.txt
drwxr-xr-x   - root supergroup          0 2017-08-26 11:19 /user/hive
drwxr-xr-x   - root supergroup          0 2017-08-29 17:48 /user/root
[root@master /]# hdfs dfs -mv /log.txt /log2.txt    --重命名
[root@master /]# hdfs dfs -ls /
Found 5 items
drwxr-xr-x   - root supergroup          0 2017-08-31 10:16 /hbase
-rw-r--r--   2 root root                0 2018-01-13 17:26 /log2.txt
drwxrwxrwx   - root supergroup          0 2018-01-13 16:36 /test
drwxrwxrwx   - root supergroup          0 2017-08-24 17:28 /tmp
drwxr-xr-x   - root supergroup          0 2018-01-13 17:34 /user
```

10. cp

复制文件，例如：

[root@master /]# hdfs dfs -cp /log2.txt /log.txt
[root@master /]# hdfs dfs -ls /

drwxr-xr-x	- root supergroup	0 2017-08-31 10:16	/hbase
-rw-r--r--	2 root supergroup	0 2018-01-13 17:37	/log.txt --新复制文件
-rw-r--r--	2 root root	0 2018-01-13 17:26	/log2.txt
drwxrwxrwx	- root supergroup	0 2018-01-13 16:36	/test
drwxrwxrwx	- root supergroup	0 2017-08-24 17:28	/tmp
drwxr-xr-x	- root supergroup	0 2018-01-13 17:34	/user

选项-f 为如果文件已经存在，则覆盖已有文件。

[root@master /]# hdfs dfs -cp /log2.txt /log.txt
cp: `/ log.txt': File exists   --文件已经存在时给出错误提示
[root@master /]# hdfs dfs -cp -f /log2.txt /log.txt

11. mkdir

创建文件夹，例如：

[root@master /]# hdfs dfs -mkdir /test2
[root@master /]# hdfs dfs -ls /
[root@master /]# hdfs dfs -ls /
Found 7 items

drwxr-xr-x	- root supergroup	0 2017-08-31 10:16	/hbase
-rw-r--r--	2 root supergroup	0 2018-01-13 17:37	/log.txt
-rw-r--r--	2 root root	0 2018-01-13 17:26	/log2.txt
drwxrwxrwx	- root supergroup	0 2018-01-13 16:36	/test
drwxr-xr-x	- root supergroup	0 2018-01-13 17:39	/test2
drwxrwxrwx	- root supergroup	0 2017-08-24 17:28	/tmp
drwxr-xr-x	- root supergroup	0 2018-01-13 17:34	/user

选项-p 为如果上层目录不存在，则递归建立所需目录。

[root@master /]# hdfs dfs -mkdir /test1/test2
mkdir: `/test1/test2': No such file or directory   --上层目录不存在，给出错误提示
[root@master /]# hdfs dfs -mkdir -p /test1/test2
[root@master /]# hdfs dfs -ls /test1
Found 1 items
drwxr-xr-x   - root supergroup   0 2018-01-13 17:40 /test1/test2

12. rm

删除文件，例如：

[root@master /]# hdfs dfs -rm /log2.txt
18/01/14 09:15:26 INFO fs.TrashPolicyDefault: Namenode trash configuration: Deletion interval = 0 minutes, Emptier interval = 0 minutes.
Deleted /log2.txt

选项-r 为递归的删除，可以删除非空目录。

[root@master /]# hdfs dfs -rm /test1
rm: `/test1': Is a directory   --删除文件夹时给出错误提示
[root@master /]# hdfs dfs -rm -r /test1   --使用-r 选项，文件夹及文件夹下的文件删除成功
18/01/14 09:17:27 INFO fs.TrashPolicyDefault: Namenode trash configuration: Deletion interval = 0 minutes, Emptier interval = 0 minutes.
Deleted /test1

[root@master data]# hdfs dfs -rm /*.txt   --删除/目录下所有.txt 结尾的文件
18/01/23 09:12:16 INFO fs.TrashPolicyDefault: Namenode trash configuration: Deletion interval = 0 minutes, Emptier interval = 0 minutes.
Deleted /1.txt
18/01/23 09:12:16 INFO fs.TrashPolicyDefault: Namenode trash configuration: Deletion interval = 0 minutes, Emptier interval = 0 minutes.
Deleted /2.txt
18/01/23 09:12:16 INFO fs.TrashPolicyDefault: Namenode trash configuration: Deletion interval = 0 minutes, Emptier interval = 0 minutes.
Deleted /log.txt

13. rmdir

删除空目录，例如：

[root@master /]# hdfs dfs -rmdir /test1
rmdir: `/test1': Directory is not empty   --不能删除非空目录

选项--ignore-fail-on-non-empty 为忽略非空删除失败时的提示。

[root@master /]# hdfs dfs -rmdir --ignore-fail-on-non-empty /test1
[root@master /]# hdfs dfs -ls /
Found 7 items
drwxr-xr-x   - root supergroup   0 2017-08-31 10:16 /hbase
-rw-r--r--   2 root supergroup   0 2018-01-13 17:37 /log.txt
drwxrwxrwx   - root supergroup   0 2018-01-13 16:36 /test
drwxr-xr-x   - root supergroup   0 2018-01-14 09:19 /test1   --不给出错误提示，但文件未删除
drwxr-xr-x   - root supergroup   0 2018-01-13 17:39 /test2
drwxrwxrwx   - root supergroup   0 2017-08-24 17:28 /tmp
drwxr-xr-x   - root supergroup   0 2018-01-13 17:34 /user

14. setrep

改变一个文件的副本数，例如：

[root@master /]# hdfs dfs -stat %r /log.txt
1   --原副本数
[root@master /]# hdfs dfs -setrep 2 /log.txt
Replication 2 set: /log.txt
[root@master /]# hdfs dfs -stat %r /log.txt
2   --改变后的副本数

选项-w 为命令等待副本数调整完成。

[root@master /]# hdfs dfs -setrep -w 1 /log.txt
Replication 1 set: /log.txt
Waiting for /log.txt ... done
[root@master /]# hdfs dfs -stat %r /log.txt
1

15. expunge

清空回收站，例如：

[root@master /]# hdfs dfs -expunge
18/01/14 09:23:25 INFO fs.TrashPolicyDefault: Namenode trash configuration: Deletion interval = 0 minutes, Emptier interval = 0 minutes.

16. chgrp

修改文件用户组，例如：

```
[root@master /]# hdfs dfs -ls /
Found 7 items
drwxr-xr-x   - root supergroup          0 2017-08-31 10:16 /hbase
-rw-r--r--   2 root supergroup          0 2018-01-13 17:37 /log.txt
drwxrwxrwx   - root supergroup          0 2018-01-13 16:36 /test
drwxr-xr-x   - root supergroup          0 2018-01-14 09:19 /test1
drwxr-xr-x   - root supergroup          0 2018-01-13 17:39 /test2
drwxrwxrwx   - root supergroup          0 2017-08-24 17:28 /tmp
drwxr-xr-x   - root supergroup          0 2018-01-13 17:34 /user
[root@master /]# hdfs dfs -chgrp test /test1
[root@master /]# hdfs dfs -ls /
Found 7 items
drwxr-xr-x   - root supergroup          0 2017-08-31 10:16 /hbase
-rw-r--r--   2 root supergroup          0 2018-01-13 17:37 /log.txt
drwxrwxrwx   - root supergroup          0 2018-01-13 16:36 /test
drwxr-xr-x   - root test                0 2018-01-14 09:19 /test1    --修改后的用户组（未建立 test 组
                                                                       --仍可成功）
drwxr-xr-x   - root supergroup          0 2018-01-13 17:39 /test2
drwxrwxrwx   - root supergroup          0 2017-08-24 17:28 /tmp
drwxr-xr-x   - root supergroup          0 2018-01-13 17:34 /user
[root@master /]# hdfs dfs -ls /test1
Found 1 items
drwxr-xr-x   - root supergroup          0 2018-01-14 09:19 /test1/test2   --目录下文件的用户组未修改
```

选项 -R 为递归修改，如果是目录，则递归修改其下的文件及目录。

```
[root@master /]# hdfs dfs -chgrp -R test /test1
[root@master /]# hdfs dfs -ls /test1
Found 1 items
drwxr-xr-x   - root test1               0 2018-01-14 09:19 /test1/test2
```

17. chmod

修改文件权限，权限模式同 Linux shell 命令中的模式，例如：

```
[root@master /]# hdfs dfs -ls /
Found 7 items
drwxr-xr-x   - root supergroup          0 2017-08-31 10:16 /hbase
-rw-r--r--   2 root supergroup          0 2018-01-13 17:37 /log.txt
drwxrwxrwx   - root supergroup          0 2018-01-13 16:36 /test
drwxr-xr-x   - root test                0 2018-01-14 09:19 /test1
drwxr-xr-x   - root supergroup          0 2018-01-13 17:39 /test2
drwxrwxrwx   - root supergroup          0 2017-08-24 17:28 /tmp
drwxr-xr-x   - root supergroup          0 2018-01-13 17:34 /user
[root@master /]# hdfs dfs -chmod 777 /test1
[root@master /]# hdfs dfs -ls /
Found 7 items
drwxr-xr-x   - root supergroup          0 2017-08-31 10:16 /hbase
```

-rw-r--r--	2 root supergroup	0 2018-01-13 17:37 /log.txt	
drwxrwxrwx	- root supergroup	0 2018-01-13 16:36 /test	
drwxrwxrwx	- root test	0 2018-01-14 09:19 /test1	--修改后的权限
drwxr-xr-x	- root supergroup	0 2018-01-13 17:39 /test2	
drwxrwxrwx	- root supergroup	0 2017-08-24 17:28 /tmp	
drwxr-xr-x	- root supergroup	0 2018-01-13 17:34 /user	

[root@master /]# hdfs dfs -ls /test1
Found 1 items

drwxr-xr-x	- root test1	0 2018-01-14 09:19 /test1/test2	--目录下文件的权限未修改

[root@master /]# hdfs dfs -chmod -R 775 /test1
[root@master /]# hdfs dfs -ls /test1
Found 1 items

drwxrwxrwx	- root test1	0 2018-01-14 09:19 /test1/test2

18. chown

修改文件的用户或组，例如：

[root@master /]# hdfs dfs -ls /
Found 7 items

drwxr-xr-x	- root supergroup	0 2017-08-31 10:16 /hbase
-rw-r--r--	2 root supergroup	0 2018-01-13 17:37 /log.txt
drwxrwxrwx	- root supergroup	0 2018-01-13 16:36 /test
drwxrwxrwx	- root test	0 2018-01-14 09:19 /test1
drwxr-xr-x	- root supergroup	0 2018-01-13 17:39 /test2
drwxrwxrwx	- root supergroup	0 2017-08-24 17:28 /tmp
drwxr-xr-x	- root supergroup	0 2018-01-13 17:34 /user

[root@master /]# hdfs dfs -chown test /test1
[root@master /]# hdfs dfs -ls /
Found 7 items

drwxr-xr-x	- root supergroup	0 2017-08-31 10:16 /hbase	
-rw-r--r--	2 root supergroup	0 2018-01-13 17:37 /log.txt	
drwxrwxrwx	- root supergroup	0 2018-01-13 16:36 /test	
drwxrwxrwx	- test test	0 2018-01-14 09:19 /test1	--修改后的用户（未建立 test --用户，仍可成功）
drwxr-xr-x	- root supergroup	0 2018-01-13 17:39 /test2	
drwxrwxrwx	- root supergroup	0 2017-08-24 17:28 /tmp	
drwxr-xr-x	- root supergroup	0 2018-01-13 17:34 /user	

[root@master /]# hdfs dfs -ls /test1
Found 1 items

drwxrwxrwx	- root test1	0 2018-01-14 09:19 /test1/test2	--目录下文件的用户未修改

选项-R 为递归修改，如果是目录，则递归修改其下的文件及目录。

[root@master /]# hdfs dfs -chown -R test:test /test1
[root@master /]# hdfs dfs -ls /test1
Found 1 items

drwxrwxrwx	- test test	0 2018-01-14 09:19 /test1/test2

19. getfacl

显示访问控制列表 ACLs（Access Control Lists），例如：

```
[hadoop@localhost bin]# hdfs dfs -getfacl /log.txt
# file: /log.txt
# owner: root
# group: supergroup
user::rw-
group::r--
other::r--
```

选项-R 为递归显示。

```
[hadoop@localhost bin]# hdfs dfs -getfacl -R /test1
# file: /test1
# owner: test
# group: test
user::rwx
group::rwx
other::rwx


# file: /test1/test2
# owner: test
# group: test
user::rwx
group::rwx
other::rwx
```

20. setfacl

设置访问控制列表，ACLs 默认未开启，直接使用该命令会报错，例如：

```
[hadoop@localhost bin]# hdfs dfs -setfacl -b /log.txt
setfacl: The ACL operation has been rejected.  Support for ACLs has been disabled by setting dfs.namenode.acls.enabled to false.
```

开启 acls，配置 hdfs-site.xml

```
[root@master /]$ vi etc/hadoop/hdfs-site.xml
<property>
    <name>dfs.namenode.acls.enabled</name>
    <value>true</value>
</property>
```

选项-m 为修改 ACLs。

```
[root@master /]# hdfs dfs -getfacl /log.txt
# file: /log.txt
# owner: root
# group: supergroup
user::rw-
group::r--
other::r--
[root@master /]# hdfs dfs -setfacl -m user::rw-,user:hadoop:rw-,group::r--,other::r-- /log.txt
[root@master /]# hdfs dfs -getfacl /log.txt
# file: /log.txt
# owner: root
# group: supergroup
```

user::rw-
user:hadoop:rw-
group::r--
mask::rw-
other::r--

选项-x 为删除指定规则。

[root@master /]# hdfs dfs -setfacl -m user::rw-,user:hadoop:rw-,group::r--,other::r-- /log.txt
[root@master /]# hdfs dfs -getfacl /log.txt
file: /log.txt
owner: root
group: supergroup
user::rw-
user:hadoop:rw-
group::r--
mask::rw-
other::r--
[root@master /]# hdfs dfs -setfacl -x user:hadoop /log.txt
[root@master /]# hdfs dfs -getfacl /log.txt
file: /log.txt
owner: root
group: supergroup
user::rw-
group::r--
mask::r--
other::r--

选项-b 为基本的 ACL 规则（所有者、群组、其他）被保留，其他规则全部删除。

选项-k 为删除默认规则。

21. setfattr

设置扩展属性的名称和值，例如：

选项-n 为属性名称。

选项-v 为属性值。

[root@master /]# hdfs dfs -getfattr -d /log.txt
file: /log.txt
[root@master /]# hdfs dfs -setfattr -n user.web -v www.baidu.com /log.txt
[root@master /]# hdfs dfs -getfattr -d /log.txt
# file: /log.txt
user.web="www.baidu.com"

选项-x 为删除扩展属性。

[root@master /]# hdfs dfs -getfattr -d /log.txt
# file: /log.txt
user.web="www.baidu.com"
[root@master /]# hdfs dfs -setfattr -x user.web /log.txt
[root@master /]# hdfs dfs -getfattr -d /log.txt
file: /log.txt

22. getfattr

显示扩展属性的名称和值，例如：

[root@master /]# hdfs dfs -getfattr -d /log.txt
# file: /log.txt
user.web="www.baidu.com"
user.web2="www.163.com"

选项-n 为显示指定名称的属性值。

[root@master /]# hdfs dfs -getfattr -n user.web /log.txt # file: /log.txt
user.web="www.baidu.com"

4.3.4 统计相关命令

1. count

显示指定文件或目录。其中 DIR_COUNT、FILE_COUNT、CONTENT_SIZE、FILE_NAME 分别表示子目录个数（如果指定路径是目录，则包含该目录本身）、文件个数、使用字节个数，以及文件或目录名。例如：

```
[root@master /]# hdfs dfs -ls /
Found 7 items
drwxr-xr-x    - root supergroup           0 2017-08-31 10:16 /hbase
-rw-r--r--    2 root supergroup           0 2018-01-13 17:37 /log.txt
drwxrwxrwx    - root supergroup           0 2018-01-13 16:36 /test
drwxrwxrwx    - test test                 0 2018-01-14 09:19 /test1
drwxr-xr-x    - root supergroup           0 2018-01-13 17:39 /test2
drwxrwxrwx    - root supergroup           0 2017-08-24 17:28 /tmp
drwxr-xr-x    - root supergroup           0 2018-01-13 17:34 /user
[root@master /]# hdfs dfs -count /
           81             58              213574 /
```

选项-q 为显示配额信息（在多人共用的情况下，可以通过限制用户写入目录并设置目录的 quota 来防止不小心就把所有的空间用完造成别人无法存取的情况）。配额信息包括 QUOTA、REMAINING_QUOTA、SPACE_QUOTA、REMAINING_SPACE_QUOTA，分别表示某个目录下文件及目录的总数、剩余目录或文件数量、目录下空间的大小、目录下的剩余空间。

计算公式：

QUOTA - (DIR_COUNT + FILE_COUNT) = REMAINING_QUOTA
SPACE_QUOTA - CONTENT_SIZE = REMAINING_SPACE_QUOTA

none 和 inf 表示未配置。

```
[root@master /]# hdfs dfs -count -q /
9223372036854775807 9223372036854775668        none      inf      81      58      213574 /
```

2. du

显示文件大小，如果指定目录，会显示该目录中每个文件的大小。例如：

```
[root@master /]# hdfs dfs -ls -R /
drwxr-xr-x    - hadoop supergroup          0 2015-03-27 19:19 /input
-rw-r--r--    1 hadoop supergroup         14 2015-03-27 19:19 /input/input1.txt
-rw-r--r--    1 hadoop supergroup         32 2015-03-27 19:19 /input/input2.txt
```

```
-rw-r--r--         1 hadoop supergroup       28 2015-04-02 07:32 /input.txt
-rwxrwxrwx         1 hadoop hadoops          28 2015-03-31 08:59 /input1.txt
-rw-r--r--         1 hadoop supergroup      184 2015-03-31 08:14 /input1.zip
drwxr-xr-x         - hadoop supergroup        0 2015-03-27 19:16 /output
drwxr-xr-x         - hadoop supergroup        0 2015-04-02 07:29 /text
[root@master /]# hdfs dfs -du /
10734       /hbase
0           /log.txt
4671        /test
0           /test1
0           /test2
198142      /tmp
27          /user
```

选项-s 为显示总的统计信息，而不是显示每个文件的信息。

```
[root@master /]# hdfs dfs -du -s /
213574    /
```

3. df

检查文件系统的磁盘空间占用情况，例如：

```
[root@master /]# hdfs dfs -df /
Filesystem                    Size        Used      Available   Use%
hdfs://master:9000      37492883456    1130496    28328345600     0%
```

4. stat

显示文件统计信息，例如：

格式：%b 为文件所占的块数，%g 为文件所属的用户组，%n 为文件名，%o 为数据块大小，%r 为备份数，%u 为文件所属用户，%y 为文件修改时间。例如：

```
[root@master /]# hdfs dfs -stat %b,%g,%n,%o,%r,%u,%y /user
0,supergroup,user,0,0,root,2018-01-13 09:34:21
```

4.3.5 快照命令

1. createSnapshot

创建快照。Snapshot（快照）是一个全部文件系统或者某个目录在某一时刻的镜像。创建动作仅仅是在目录对应的 inode 上加个快照的标签，不会涉及数据块的拷贝操作，也不会对读写性能有影响，但是会占用 NameNode 一定的额外内存来存放快照中被修改的文件和目录的元信息。例如：

```
[root@master /]# hdfs dfs -ls -R /user
-rw-r--r--    2 root supergroup    0 2018-01-13 17:16 /user/1.txt
drwxr-xr-x    - root supergroup    0 2017-08-26 11:19 /user/hive
drwxr-xr-x    - root supergroup    0 2017-08-26 11:19 /user/hive/warehouse
drwxr-xr-x    - root supergroup    0 2018-01-13 17:11 /user/hive/warehouse/t1
-rw-r--r--    2 root supergroup   27 2017-08-26 11:29 /user/hive/warehouse/t1/t1.txt
-rw-r--r--    2 root supergroup    0 2018-01-13 17:11 /user/hive/warehouse/t1/t2.txt
drwxr-xr-x    - root supergroup    0 2017-08-29 17:48 /user/root
drwxr-xr-x    - root supergroup    0 2017-08-29 18:05 /user/root/.sparkStaging
[root@master /]# hdfs dfs -createSnapshot /user snap1
```

```
createSnapshot: Directory is not a snapshottable directory: /user      --直接创建给出错误
[root@master /]# hdfs dfsadmin -allowSnapshot /user    --对开启某一目录的快照功能
Allowing snaphot on /output succeeded
[root@master /]# hdfs dfs -createSnapshot /user snap1    --创建快照
Created snapshot /user/.snapshot/snap1
[root@master /]# hdfs dfs -ls -R /user/.snapshot
drwxr-xr-x   - root supergroup          0 2018-01-14 09:53 /user/.snapshot/snap1
-rw-r--r--   2 root supergroup          0 2018-01-13 17:16 /user/.snapshot/snap1/1.txt
drwxr-xr-x   - root supergroup          0 2017-08-26 11:19 /user/.snapshot/snap1/hive
drwxr-xr-x   - root supergroup          0 2017-08-26 11:19 /user/.snapshot/snap1/hive/warehouse
drwxr-xr-x   - root supergroup          0 2018-01-13 17:11 /user/.snapshot/snap1/hive/warehouse/t1
-rw-r--r--   2 root supergroup         27 2017-08-26 11:29 /user/.snapshot/snap1/hive/warehouse/t1/t1.txt
-rw-r--r--   2 root supergroup          0 2018-01-13 17:11 /user/.snapshot/snap1/hive/warehouse/t1/t2.txt
drwxr-xr-x   - root supergroup          0 2017-08-29 17:48 /user/.snapshot/snap1/root
drwxr-xr-x   - root supergroup          0 2017-08-29 18:05 /user/.snapshot/snap1/root/.sparkStaging
```

2. renameSnapshot

重命名快照，例如：

```
[root@master /]# hdfs dfs -ls /user/.snapshot/snap1
Found 3 items
-rw-r--r--   2 root supergroup          0 2018-01-13 17:16 /user/.snapshot/snap1/1.txt
drwxr-xr-x   - root supergroup          0 2017-08-26 11:19 /user/.snapshot/snap1/hive
drwxr-xr-x   - root supergroup          0 2017-08-29 17:48 /user/.snapshot/snap1/root
```

3. deleteSnapshot

删除快照，例如：

```
[root@master /]# hdfs dfs -deleteSnapshot /user snap1
[root@master /]# hdfs dfs -ls /user/.snapshot/snap1
ls: `/user/.snapshot/snap1': No such file or directory
```

4.4　本章小结

　　HDFS 是存储大数据文件的重要载体，具有高容错性、流式文件写入、可构建在廉价机器上等特点，但是它也有低延时数据访问、不支持并发写入、文件随机修改等不足。NameNode 负责管理整个 HDFS 的元数据，DataNode 负责管理用户的文件数据块，再通过 Client，即可完成 HDFS 读写操作。HDFS shell 是一种类似 Linux shell 的命令，可以方便用户操作目录和文件。

第 5 章　HDFS Java API 编程

　　HDFS 不仅可以通过 shell 操作 Hadoop 集群中的文件系统，还提供 Java API 接口完成相应的功能。本章会对 HDFS Java API 编程进行详细讲解，同时为了方便开发调试，本书中所有的 Java API 开发都采用 Windows 远程开发，因此需要在 Windows 中搭建 Hadoop 的开发环境，如果是在 Linux 下开发，则无需这些操作。

5.1　远程开发环境搭建

1. 将 hadoop-2.6.5.tar.gz 解压到 Windows 下

　　在解压过程中可能会有错误提示，这是由于 Linux 文件格式和 Windows 不一样造成的，可以忽略。

2. 下载 Eclipse

　　Eclipse 的下载网址为http://www.eclipse.org/downloads/packages/release/Mars/1，如图 5-1 所示。

图 5-1　Eclipse 下载页面

　　下载后解压即可。远程开发需要使用插件 hadoop-eclipse-2.6.0.jar（2.6.0 表示对应 Hadoop 版本号，需保持一致），下载地址为https://github.com/winghc/hadoop2x-eclipse-plugin，将下载好的插件放入 eclipse 安装目录下的 dropins 文件夹中，如图 5-2 所示。

图 5-2　插件存放位置

　　放置好后，按照图 5-3 中的步骤进行操作。

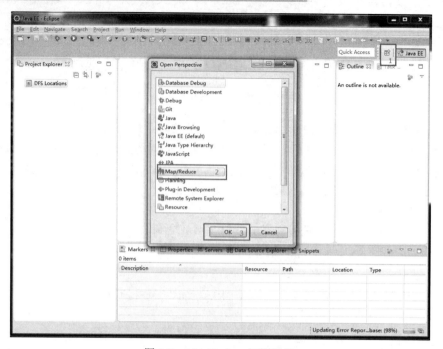

图 5-3　Map/Reduce 视图设置

单击 Map/Reduce Locations，在其下空白处右击并选择 New Hadoop Location 选项，如图 5-4 所示。

图 5-4　Hadoop Location 设置 1

将值修改为如图 5-5 所示，1 是为该配置设置一个名字，2 是填写整个 Hadoop 集群 master 的 IP 地址，3、4 分别设置通信端口号，5 为用户名，设置完毕后单击 Finish 按钮。

单击左边的 DFS Locations 即可看到如图 5-6 所示的界面，由于整个集群并没有进行任何操作，只有 tmp 和 user 两个文件夹。

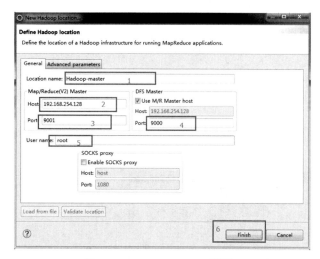

图 5-5　Hadoop Location 设置 2

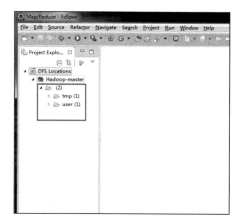

图 5-6　HDFS 视图

该方法为 HDFS 的可视化操作，可以通过右击方式对其执行创建文件夹和上传下载文件操作，如图 5-7 所示。

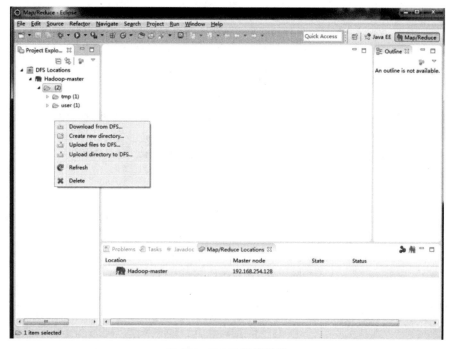

图 5-7　HDFS 视图操作

此时可能会出现如图 5-8 所示的权限错误，可以通过 HDFS shell 更改权限（hdfs dfs -chmod 777 /tmp）并更新，如图 5-9 所示。

还可以通过访问地址 http://192.168.254.128:50070/ 查看 HDFS，如图 5-10 和图 5-11 所示。

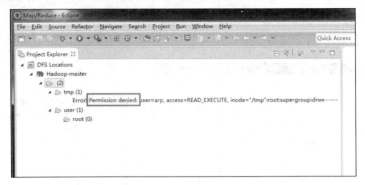

图 5-8　权限错误

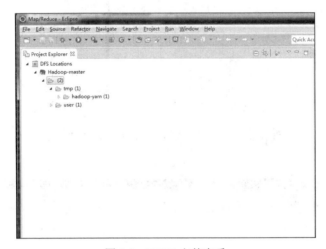

图 5-9　HDFS 文件查看

图 5-10　HDFS Web 访问接口

图 5-11　Web 方式查看 HDFS

5.2　HDFS Java API 接口

Java 抽象类 org.apache.hadoop.fs.FileSystem 定义了 Hadoop 的一个文件系统接口。该类是一个抽象类，通过以下两种静态工厂方法可以实现 FileSystem 实例：

  public static FileSystem.get(Configuration conf) throws IOException
  public static FileSystem.get(URI uri, Configuration conf) throws IOException

（1）public boolean mkdirs(Path f) throws IOException：一次性新建所有目录（包括父目录），f 是完整的目录路径。

（2）public FSOutputStream create(Path f) throws IOException：创建指定 Path 对象的一个文件，返回一个用于写入数据的输出流。create()有多个重载版本，允许我们指定是否强制覆盖已有的文件、文件备份数量、写入文件缓冲区大小、数据块大小和文件权限。

（3）public boolean copyFromLocal(Path src, Path dst) throws IOException：将本地文件拷贝到文件系统。

（4）public boolean exists(Path f) throws IOException：检查文件或目录是否存在。

（5）public boolean delete(Path f, Boolean recursive)：永久性删除指定的文件或目录，如果 f 是一个空目录或者文件，那么 recursive 的值就会被忽略。只有 recursive 为 true 时，一个非空目录及其内容才会被删除。

（6）FileStatus 类封装了文件系统中文件和目录的元数据，包括文件长度、块大小、备份、修改时间、所有者和权限信息。通过 FileStatus.getPath()可以查看指定 HDFS 中某个目录下的所有文件。

5.3　HDFS Java API 编程

1. 创建 Map/Reduce 工程

HDFS Java API 编程需要通过 Eclipse 创建 Map/Reduce 工程，如图 5-12 所示。

然后指定 Hadoop 在 Windows 下的解压目录，如图 5-13 和图 5-14 所示。

2. Java API 开发

Hadoop 通过 Configuration 类来保存配置信息，这些信息都以 XML 文件形式保存。通过 new Configuration()实例化一个对象即可获取相应信息，但是如果采用 Windows 远程开发，还需要通过以下三种方式来设置 HDFS Configuration 信息：

（1）通过 set 函数设置。
  Configuration conf = new Configuration();
  conf.set("fs.defaultFS", "hdfs://192.168.254.128:9000");

（2）通过加载配置文件。
  Configuration conf = new Configuration();
  conf.addResource(new Path("C:\\hadoop\\core-site.xml"));

core-site.xml 文件中的内容为 3.2 节配置的 core-site.xml。

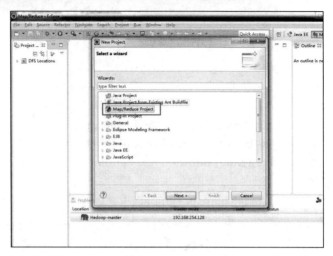

图 5-12　创建 HDFS 工程

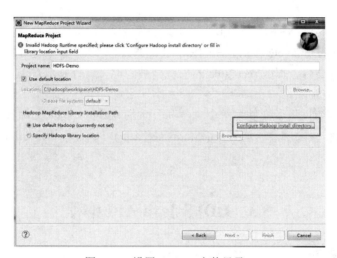

图 5-13　设置 Hadoop 安装目录 1

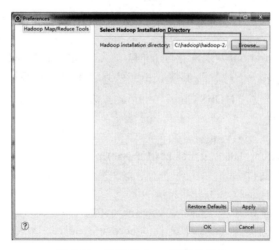

图 5-14　设置 Hadoop 安装目录 2

（3）通过设置 URI 的方式。
　　Configuration conf = new Configuration();
　　FileSystem fs = FileSystem.get(new URI("hdfs://192.168.254.128:9000"), conf);

5.3.1　获取文件系统

由于是远程访问 HDFS 文件系统，当前操作系统不是超级管理员，不具备文件的写入权限，运行会报如图 5-15 所示的异常。

```
log4j:WARN No appenders could be found for logger (org.apache.hadoop.metrics2.lib.MutableMetricsFactory).
log4j:WARN Please initialize the log4j system properly.
log4j:WARN See http://logging.apache.org/log4j/1.2/faq.html#noconfig for more info.
org.apache.hadoop.ipc.RemoteException(org.apache.hadoop.security.AccessControlException): Access denied for
    at org.apache.hadoop.hdfs.server.namenode.FSPermissionChecker.checkSuperuserPrivilege(FSPermissionCh
    at org.apache.hadoop.hdfs.server.namenode.FSNamesystem.checkSuperuserPrivilege(FSNamesystem.java:651
    at org.apache.hadoop.hdfs.server.namenode.FSNamesystem.datanodeReport(FSNamesystem.java:5488)
    at org.apache.hadoop.hdfs.server.namenode.NameNodeRpcServer.getDatanodeReport(NameNodeRpcServer.java
    at org.apache.hadoop.hdfs.protocolPB.ClientNamenodeProtocolServerSideTranslatorPB.getDatanodeReport(
    at org.apache.hadoop.hdfs.protocol.proto.ClientNamenodeProtocolProtos$ClientNamenodeProtocol$2.callB
    at org.apache.hadoop.ipc.ProtobufRpcEngine$Server$ProtoBufRpcInvoker.call(ProtobufRpcEngine.java:619
    at org.apache.hadoop.ipc.RPC$Server.call(RPC.java:975)
    at org.apache.hadoop.ipc.Server$Handler$1.run(Server.java:2040)
    at org.apache.hadoop.ipc.Server$Handler$1.run(Server.java:2036)
    at java.security.AccessController.doPrivileged(Native Method)
    at javax.security.auth.Subject.doAs(Subject.java:422)
    at org.apache.hadoop.security.UserGroupInformation.doAs(UserGroupInformation.java:1692)
    at org.apache.hadoop.ipc.Server$Handler.run(Server.java:2034)

    at org.apache.hadoop.ipc.Client.call(Client.java:1470)
    at org.apache.hadoop.ipc.Client.call(Client.java:1401)
    at org.apache.hadoop.ipc.ProtobufRpcEngine$Invoker.invoke(ProtobufRpcEngine.java:232)
    at com.sun.proxy.$Proxy9.getDatanodeReport(Unknown Source)
    at org.apache.hadoop.hdfs.protocolPB.ClientNamenodeProtocolTranslatorPB.getDatanodeReport(ClientName
    at sun.reflect.NativeMethodAccessorImpl.invoke0(Native Method)
    at sun.reflect.NativeMethodAccessorImpl.invoke(Unknown Source)
    at sun.reflect.DelegatingMethodAccessorImpl.invoke(Unknown Source)
    at java.lang.reflect.Method.invoke(Unknown Source)
    at org.apache.hadoop.io.retry.RetryInvocationHandler.invokeMethod(RetryInvocationHandler.java:187)
    at org.apache.hadoop.io.retry.RetryInvocationHandler.invoke(RetryInvocationHandler.java:102)
    at com.sun.proxy.$Proxy10.getDatanodeReport(Unknown Source)
    at org.apache.hadoop.hdfs.DFSClient.datanodeReport(DFSClient.java:2390)
    at org.apache.hadoop.hdfs.DistributedFileSystem.getDataNodeStats(DistributedFileSystem.java:1009)
    at org.apache.hadoop.hdfs.DistributedFileSystem.getDataNodeStats(DistributedFileSystem.java:1003)
    at Test1.listDataNodeInfo(Test1.java:49)
```

图 5-15　程序异常

出现这个异常的原因是 Administrator 用户没有权限，在代码中添加如下信息即可，其中 "root"是有访问 Hadoop 集群权限的用户名：System.setProperty("HADOOP_USER_NAME", "root")。

获取文件系统代码如下：
　　/**
　　　* 获取文件系统
　　　*
　　　* @return FileSystem
　　　*/
　　public static FileSystem getFileSystem() {
　　　　//读取配置文件
　　　　Configuration conf = new Configuration();
　　　　// 文件系统
　　　　FileSystem fs = null;

```
        conf.set("fs.defaultFS", "hdfs://192.168.254.128:9000");
            System.setProperty("HADOOP_USER_NAME", "root");
            try {
                fs = FileSystem.get(conf);
            } catch (IOException e) {
                // TODO Auto-generated catch block
                e.printStackTrace();
            }
            return fs;
    }
```

5.3.2 列出所有 DataNode 的名字信息

由于 DataNode 是分布在不同节点，因此需要通过 DistributedFileSystem 类进行访问，该类包含了 DataNode 很多有用信息，通过 DatanodeInfo.getHostName()可获取 HDFS 集群上的所有节点名称。具体实现如下：

```
    /**
     * 列出所有 DataNode 的名字信息
     */
    public static void listDataNodeInfo() {
        try {
          Configuration conf = new Configuration();
          conf.set("fs.defaultFS", "hdfs://192.168.254.128:9000");
          System.setProperty("HADOOP_USER_NAME", "root");
            FileSystem fs = FileSystem.get(conf);
            DistributedFileSystem hdfs = (DistributedFileSystem)fs;
            DatanodeInfo[] dataNodeStats = hdfs.getDataNodeStats();
            String[] names = new String[dataNodeStats.length];
            System.out.println("List of all the datanodes in the HDFS cluster:");

            for (int I = 0; I < names.length; i++) {
                names[i] = dataNodeStats[i].getHostName();
                System.out.println(names[i]);
            }
            System.out.println(hdfs.getUri().toString());
        } catch (Exception e) {
            e.printStackTrace();
        }
    }
```

测试代码为：

```
    public static void main(String[] args) throws IOException, URISyntaxException {
        listDataNodeInfo();
    }
```

运行结果如图 5-16 所示。

```
List of all the datanodes in the HDFS cluster:
slave2
slave1
hdfs://192.168.254.128:9000
```

图 5-16　HDFS 中 DataNode 的名字信息

5.3.3　创建文件目录

创建文件目录通过 mkdirs 方法实现，该方法会返回一个布尔值，表示创建文件夹是否成功。具体代码如下：

```
/**
 * 创建文件目录
 * @param path：创建的目录名
 */
public static void mkdir(String path) throws IOException{
    Configuration conf = new Configuration();
    conf.set("fs.defaultFS", "hdfs://192.168.254.128:9000");
    System.setProperty("HADOOP_USER_NAME", "root");
    FileSystem fs = FileSystem.get(conf);
    Path srcPath = new Path(path);
    boolean isok = fs.mkdirs(srcPath);
    if(isok){
        System.out.println("create dir ok!");
    }else{
        System.out.println("create dir failure");
    }
    fs.close();
}
```

测试代码为：

```
public static void main(String[] args) throws IOException, URISyntaxException {
    mkdir("/test2");
}
```

运行结果通过 HDFS shell 可以查看，如图 5-17 所示。

```
[root@master ~]# hdfs dfs -ls /
Found 4 items
drwxr-xr-x   - root supergroup          0 2017-08-31 10:16 /hbase
drwxrwxrwx   - root supergroup          0 2017-08-24 17:24 /test
drwxrwxrwx   - root supergroup          0 2017-08-24 17:28 /tmp
drwxr-xr-x   - root supergroup          0 2017-08-26 11:19 /user
[root@master ~]# hdfs dfs -ls /
Found 5 items
drwxr-xr-x   - root supergroup          0 2017-08-31 10:16 /hbase
drwxrwxrwx   - root supergroup          0 2017-08-24 17:24 /test
drwxr-xr-x   - root supergroup          0 2018-01-13 15:51 /test2
drwxrwxrwx   - root supergroup          0 2017-08-24 17:28 /tmp
drwxr-xr-x   - root supergroup          0 2017-08-26 11:19 /user
```

图 5-17　创建/test2 目录

在运行的时候可能会遇到如图 5-18 所示的错误。

```
log4j:WARN See http://logging.apache.org/log4j/1.2/faq.html#noconfig for more info.
Exception in thread "main" org.apache.hadoop.security.AccessControlException: Permission denied: user=arp, access=WRITE, inode="/test":root:supergroup:drwxr-xr-x
    at org.apache.hadoop.hdfs.server.namenode.FSPermissionChecker.checkFsPermission(FSPermissionChecker.java:271)
    at org.apache.hadoop.hdfs.server.namenode.FSPermissionChecker.check(FSPermissionChecker.java:257)
    at org.apache.hadoop.hdfs.server.namenode.FSPermissionChecker.check(FSPermissionChecker.java:238)
    at org.apache.hadoop.hdfs.server.namenode.FSPermissionChecker.checkPermission(FSPermissionChecker.java:179)
    at org.apache.hadoop.hdfs.server.namenode.FSNamesystem.checkPermission(FSNamesystem.java:6547)
    at org.apache.hadoop.hdfs.server.namenode.FSNamesystem.checkPermission(FSNamesystem.java:6529)
    at org.apache.hadoop.hdfs.server.namenode.FSNamesystem.checkAncestorAccess(FSNamesystem.java:6481)
    at org.apache.hadoop.hdfs.server.namenode.FSNamesystem.startFileInternal(FSNamesystem.java:2712)
    at org.apache.hadoop.hdfs.server.namenode.FSNamesystem.startFileInt(FSNamesystem.java:2632)
    at org.apache.hadoop.hdfs.server.namenode.FSNamesystem.startFile(FSNamesystem.java:2520)
    at org.apache.hadoop.hdfs.server.namenode.NameNodeRpcServer.create(NameNodeRpcServer.java:579)
    at org.apache.hadoop.hdfs.protocolPB.ClientNamenodeProtocolServerSideTranslatorPB.create(ClientNamenodeProtocolServerSideTranslatorPB.java:394)
    at org.apache.hadoop.hdfs.protocol.proto.ClientNamenodeProtocolProtos$ClientNamenodeProtocol$2.callBlockingMethod(ClientNamenodeProtocolProtos.java)
    at org.apache.hadoop.ipc.ProtobufRpcEngine$Server$ProtoBufRpcInvoker.call(ProtobufRpcEngine.java:619)
    at org.apache.hadoop.ipc.RPC$Server.call(RPC.java:975)
    at org.apache.hadoop.ipc.Server$Handler$1.run(Server.java:2040)
    at org.apache.hadoop.ipc.Server$Handler$1.run(Server.java:2036)
```

图 5-18 HDFS Java API 运行错误

这是没有访问权限造成的，可以通过 HDFS shell chmod 增加相应权限，也可以修改 Hadoop 集群中的 core-site.xml 文件新增 dfs.permissions 属性和 false 值。

5.3.4 删除文件或文件目录

删除文件或文件目录可以通过 deleteOnExit 方法实现，该方法会返回一个布尔值，表示创建文件夹是否成功。具体代码如下：

```
/**
 * 删除文件或文件目录
 * @param filePath：删除的文件或目录名
 */
public static void delete(String filePath) throws IOException{
    Configuration conf = new Configuration();
    conf.set("fs.defaultFS", "hdfs://192.168.254.128:9000");
    System.setProperty("HADOOP_USER_NAME", "root");
    FileSystem fs = FileSystem.get(conf);
    Path path = new Path(filePath);
    boolean isok = fs.deleteOnExit(path);
    if(isok){
        System.out.println("delete ok!");
    }else{
        System.out.println("delete failure");
    }
    fs.close();
}
```

测试代码为：

```
public static void main(String[] args) throws IOException, URISyntaxException {
    delete("/test2");
}
```

运行结果通过 HDFS shell 可以查看，如图 5-19 所示。

```
[root@master ~]# hdfs dfs -ls /
Found 5 items
drwxr-xr-x   - root supergroup          0 2017-08-31 10:16 /hbase
drwxrwxrwx   - root supergroup          0 2017-08-24 17:24 /test
drwxr-xr-x   - root supergroup          0 2018-01-13 15:51 /test2
drwxrwxrwx   - root supergroup          0 2017-08-24 17:28 /tmp
drwxr-xr-x   - root supergroup          0 2017-08-26 11:19 /user
[root@master ~]# hdfs dfs -ls /
Found 4 items
drwxr-xr-x   - root supergroup          0 2017-08-31 10:16 /hbase
drwxrwxrwx   - root supergroup          0 2017-08-24 17:24 /test
drwxrwxrwx   - root supergroup          0 2017-08-24 17:28 /tmp
drwxr-xr-x   - root supergroup          0 2017-08-26 11:19 /user
```

图 5-19 删除/test2 目录

5.3.5 查看文件是否存在

判断一个文件或文件目录是否存在可以通过 exists 方法实现，该方法会返回一个布尔值，为 true 表示存在，false 表示不存在。具体代码如下：

```java
/**
 * 判断文件是否存在
 * @param path：文件名
 */
public static void checkFileExist(Path path) {
    try {
        Configuration conf = new Configuration();
        conf.set("fs.defaultFS", "hdfs://192.168.254.128:9000");
        System.setProperty("HADOOP_USER_NAME", "root");
        FileSystem fs = FileSystem.get( conf);
        DistributedFileSystem hdfs = (DistributedFileSystem)fs;
        Path f = hdfs.getHomeDirectory();
        System.out.println("main path:" + f.toString());
        boolean exist = fs.exists(path);
        System.out.println("Whether exist of this file:"+exist);
    } catch (Exception e) {
        e.printStackTrace();
    }
}
```

测试代码为：

```java
public static void main(String[] args) throws IllegalArgumentException, Exception {
    checkFileExist(new Path("/test"));
}
```

运行结果如图 5-20 所示。

```
main path:hdfs://192.168.254.128:9000/user/root
Whether exist of this file:true
```

图 5-20 /test 文件存在

5.3.6 文件上传至 HDFS

copyFromLocalFile 表示文件的复制，该方法有三个参数：第一个参数是布尔值，表示是否删除源文件，true 为删除，默认为 false 不删除；第二个参数是要复制的文件；第三个参数是要复制到的目的地。

```java
/**
 * 上传文件
 * @param src：源文件名
 * @param src：目的地文件名
 */
public static void uploadFile(String src,String dst) throws IOException, URISyntaxException{
    Configuration conf = new Configuration();
```

```
        FileSystem fs = FileSystem.get(new URI("hdfs://192.168.254.128:9000"), conf);
//      FileSystem fs = FileSystem.get(conf);
        Path srcPath = new Path(src);        //源路径
        Path dstPath = new Path(dst);        //目标路径
        fs.copyFromLocalFile(false, srcPath, dstPath);
        fs.close();
    }
```

测试代码为：

```
    public static void main(String[] args) throws IllegalArgumentException, Exception {
            //测试上传文件
            uploadFile("C:\\log.txt", "/test");
    }
```

运行结果如图 5-21 所示。

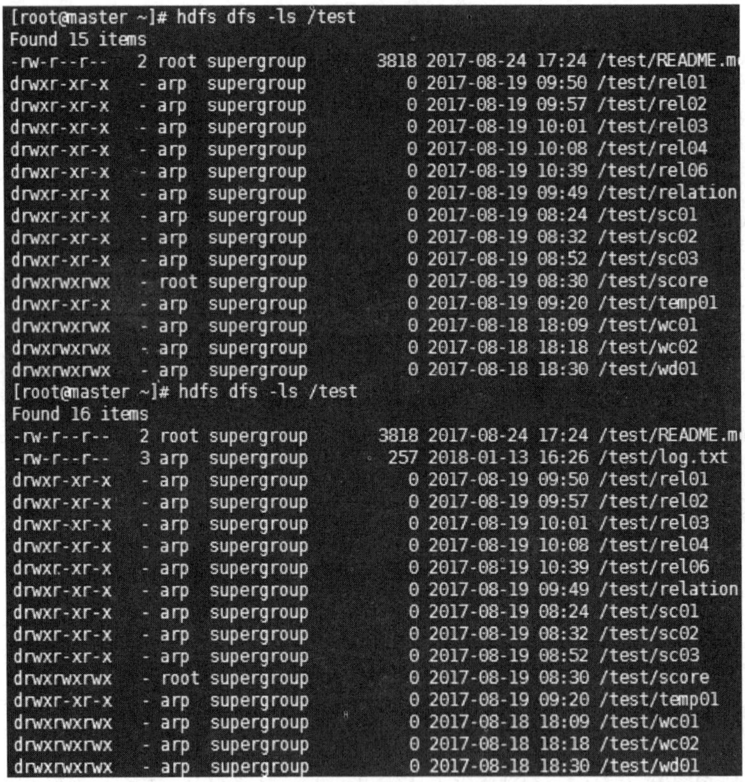

```
[root@master ~]# hdfs dfs -ls /test
Found 15 items
-rw-r--r--   2 root supergroup       3818 2017-08-24 17:24 /test/README.m
drwxr-xr-x   - arp  supergroup          0 2017-08-19 09:50 /test/rel01
drwxr-xr-x   - arp  supergroup          0 2017-08-19 09:57 /test/rel02
drwxr-xr-x   - arp  supergroup          0 2017-08-19 10:01 /test/rel03
drwxr-xr-x   - arp  supergroup          0 2017-08-19 10:08 /test/rel04
drwxr-xr-x   - arp  supergroup          0 2017-08-19 10:39 /test/rel06
drwxr-xr-x   - arp  supergroup          0 2017-08-19 09:49 /test/relation
drwxr-xr-x   - arp  supergroup          0 2017-08-19 08:24 /test/sc01
drwxr-xr-x   - arp  supergroup          0 2017-08-19 08:32 /test/sc02
drwxr-xr-x   - arp  supergroup          0 2017-08-19 08:52 /test/sc03
drwxrwxrwx   - root supergroup          0 2017-08-19 08:30 /test/score
drwxr-xr-x   - arp  supergroup          0 2017-08-19 09:20 /test/temp01
drwxrwxrwx   - arp  supergroup          0 2017-08-18 18:09 /test/wc01
drwxrwxrwx   - arp  supergroup          0 2017-08-18 18:18 /test/wc02
drwxrwxrwx   - arp  supergroup          0 2017-08-18 18:30 /test/wd01
[root@master ~]# hdfs dfs -ls /test
Found 16 items
-rw-r--r--   2 root supergroup       3818 2017-08-24 17:24 /test/README.m
-rw-r--r--   3 arp  supergroup        257 2018-01-13 16:26 /test/log.txt
drwxr-xr-x   - arp  supergroup          0 2017-08-19 09:50 /test/rel01
drwxr-xr-x   - arp  supergroup          0 2017-08-19 09:57 /test/rel02
drwxr-xr-x   - arp  supergroup          0 2017-08-19 10:01 /test/rel03
drwxr-xr-x   - arp  supergroup          0 2017-08-19 10:08 /test/rel04
drwxr-xr-x   - arp  supergroup          0 2017-08-19 10:39 /test/rel06
drwxr-xr-x   - arp  supergroup          0 2017-08-19 09:49 /test/relation
drwxr-xr-x   - arp  supergroup          0 2017-08-19 08:24 /test/sc01
drwxr-xr-x   - arp  supergroup          0 2017-08-19 08:32 /test/sc02
drwxr-xr-x   - arp  supergroup          0 2017-08-19 08:52 /test/sc03
drwxrwxrwx   - root supergroup          0 2017-08-19 08:30 /test/score
drwxr-xr-x   - arp  supergroup          0 2017-08-19 09:20 /test/temp01
drwxrwxrwx   - arp  supergroup          0 2017-08-18 18:09 /test/wc01
drwxrwxrwx   - arp  supergroup          0 2017-08-18 18:18 /test/wc02
drwxrwxrwx   - arp  supergroup          0 2017-08-18 18:30 /test/wd01
```

图 5-21　log.txt 成功上传到/test 目录

5.3.7　从 HDFS 下载文件

copyToLocalFile 表示从 HDFS 下载文件到本地系统。该方法有两个参数：第一个参数表示要下载的 HDFS 文件，第二个参数表示要下载到的目录或文件。具体代码如下：

```
    /**
     * 从 HDFS 下载文件
     * @param remote：HDFS 文件
```

```
 * @param local：目的地
 */
public static void download(String remote, String local) throws IOException, URISyntaxException {
    Configuration conf = new Configuration();
    Path path = new Path(remote);
    FileSystem fs = FileSystem.get(new URI("hdfs://192.168.254.128:9000"), conf);
    fs.copyToLocalFile(path, new Path(local));
    System.out.println("download: from" + remote + " to " + local);
    fs.close();
}
```

测试代码为：

```
public static void main(String[] args) throws IllegalArgumentException, Exception {
    download("/test/log.txt", "E:\\");
}
```

运行结果如图 5-22 所示。

```
<terminated> Test1 [Java Application] D:\Program Files\Java\jre1.8.0_144\bin\javaw.exe (2018年1月13日 下午4:31:52
log4j:WARN No appenders could be found for logger (org.apache
log4j:WARN Please initialize the log4j system properly.
log4j:WARN See http://logging.apache.org/log4j/1.2/faq.html#r
download: from/test/log.txt to E:\
```

图 5-22　文件下载成功

5.3.8　文件重命名

文件重命名可以使用 rename 方法。该方法有两个参数：第一个参数表示要重命名的文件，第二个参数表示文件的新名字。该方法返回一个布尔值，表示是否成功。具体代码如下：

```
/**
 * 文件重命名
 * @param oldName：旧文件名
 * @param newName：新文件名
 */
public static void rename(String oldName,String newName) throws IOException{
    Configuration conf = new Configuration();
    conf.set("fs.defaultFS", "hdfs://192.168.254.128:9000");
    System.setProperty("HADOOP_USER_NAME", "root");
    FileSystem fs = FileSystem.get(conf);
    Path oldPath = new Path(oldName);
    Path newPath = new Path(newName);
    boolean isok = fs.rename(oldPath, newPath);
    if(isok){
        System.out.println("rename ok!");
    }else{
        System.out.println("rename failure");
    }
}
```

```
        fs.close();
    }
```

测试代码为：

```
    public static void main(String[] args) throws IllegalArgumentException, Exception {
            rename("/test/log.txt", "/test/log2.txt");
    }
```

运行结果如图 5-23 所示。

```
[root@master ~]# hdfs dfs -ls /test
Found 16 items
-rw-r--r--   2 root supergroup       3818 2017-08-24 17:24 /test/README.md
-rw-r--r--   3 arp  supergroup        257 2018-01-13 16:26 /test/log.txt
drwxr-xr-x   - arp  supergroup          0 2017-08-19 09:50 /test/rel01
drwxr-xr-x   - arp  supergroup          0 2017-08-19 09:57 /test/rel02
drwxr-xr-x   - arp  supergroup          0 2017-08-19 10:01 /test/rel03
drwxr-xr-x   - arp  supergroup          0 2017-08-19 10:08 /test/rel04
drwxr-xr-x   - arp  supergroup          0 2017-08-19 10:39 /test/rel06
drwxr-xr-x   - arp  supergroup          0 2017-08-19 09:49 /test/relation
drwxr-xr-x   - arp  supergroup          0 2017-08-19 08:24 /test/sc01
drwxr-xr-x   - arp  supergroup          0 2017-08-19 08:32 /test/sc02
drwxr-xr-x   - arp  supergroup          0 2017-08-19 08:52 /test/sc03
drwxrwxrwx   - root supergroup          0 2017-08-19 08:30 /test/score
drwxr-xr-x   - arp  supergroup          0 2017-08-19 09:20 /test/temp01
drwxrwxrwx   - arp  supergroup          0 2017-08-18 18:09 /test/wc01
drwxrwxrwx   - arp  supergroup          0 2017-08-18 18:18 /test/wc02
drwxrwxrwx   - arp  supergroup          0 2017-08-18 18:30 /test/wd01
[root@master ~]# hdfs dfs -ls /test
Found 16 items
-rw-r--r--   2 root supergroup       3818 2017-08-24 17:24 /test/README.md
-rw-r--r--   3 arp  supergroup        257 2018-01-13 16:26 /test/log2.txt
drwxr-xr-x   - arp  supergroup          0 2017-08-19 09:50 /test/rel01
drwxr-xr-x   - arp  supergroup          0 2017-08-19 09:57 /test/rel02
drwxr-xr-x   - arp  supergroup          0 2017-08-19 10:01 /test/rel03
drwxr-xr-x   - arp  supergroup          0 2017-08-19 10:08 /test/rel04
drwxr-xr-x   - arp  supergroup          0 2017-08-19 10:39 /test/rel06
drwxr-xr-x   - arp  supergroup          0 2017-08-19 09:49 /test/relation
drwxr-xr-x   - arp  supergroup          0 2017-08-19 08:24 /test/sc01
drwxr-xr-x   - arp  supergroup          0 2017-08-19 08:32 /test/sc02
drwxr-xr-x   - arp  supergroup          0 2017-08-19 08:52 /test/sc03
drwxrwxrwx   - root supergroup          0 2017-08-19 08:30 /test/score
drwxr-xr-x   - arp  supergroup          0 2017-08-19 09:20 /test/temp01
drwxrwxrwx   - arp  supergroup          0 2017-08-18 18:09 /test/wc01
drwxrwxrwx   - arp  supergroup          0 2017-08-18 18:18 /test/wc02
drwxrwxrwx   - arp  supergroup          0 2017-08-18 18:30 /test/wd01
```

图 5-23　log.txt 重命名为 log2.txt

5.3.9　遍历目录和文件

FileStatus 包含了 HDFS 文件的常用信息，而该对象的使用需要通过 DistributedFileSystem 对象来访问，因此需要先将 FileSystem 对象强制转换为 DistributedFileSystem。具体代码如下：

```
    /**
    * 遍历文件和目录
    * @param path：文件路径
    */
    private static void showDir(Path path) throws Exception {
        Configuration conf = new Configuration();
        conf.set("fs.defaultFS", "hdfs://192.168.254.128:9000");
        System.setProperty("HADOOP_USER_NAME", "root");
        FileSystem fs = FileSystem.get(conf);
        DistributedFileSystem hdfs = (DistributedFileSystem)fs;
        FileStatus[] fileStatus = hdfs.listStatus(path);
        if(fileStatus.length > 0){
```

```java
            for(FileStatus status : fileStatus){
                Path f = status.getPath();
                System.out.println(f);
                //判断是否为目录，是的话循环遍历嵌套访问
                if(status.isDirectory()){
                    FileStatus[] files = hdfs.listStatus(f);
                    if(files.length > 0){
                        for(FileStatus file : files)
                            showDir(file.getPath());
                    }
                }
            }
        }
```

测试代码为：
```java
    public static void main(String[] args) throws IllegalArgumentException, Exception {
        showDir(new Path("/"));
    }
```

运行结果如图 5-24 所示。

```
hdfs://192.168.254.128:9000/test/rel06/_SUCCESS
hdfs://192.168.254.128:9000/test/rel06/part-r-00000
hdfs://192.168.254.128:9000/test/relation/1.txt
hdfs://192.168.254.128:9000/test/relation/2.txt
hdfs://192.168.254.128:9000/test/sc01/_SUCCESS
hdfs://192.168.254.128:9000/test/sc01/part-r-00000
hdfs://192.168.254.128:9000/test/sc02/_SUCCESS
hdfs://192.168.254.128:9000/test/sc02/part-r-00000
hdfs://192.168.254.128:9000/test/sc03/_SUCCESS
hdfs://192.168.254.128:9000/test/sc03/part-r-00000
hdfs://192.168.254.128:9000/test/score/bigdata
hdfs://192.168.254.128:9000/test/score/china
hdfs://192.168.254.128:9000/test/score/math
hdfs://192.168.254.128:9000/test/temp01/_SUCCESS
hdfs://192.168.254.128:9000/test/temp01/part-r-00000
hdfs://192.168.254.128:9000/test/wc01/_SUCCESS
hdfs://192.168.254.128:9000/test/wc01/part-r-00000
hdfs://192.168.254.128:9000/test/wc02/_SUCCESS
hdfs://192.168.254.128:9000/test/wc02/part-r-00000
hdfs://192.168.254.128:9000/test/wd01/_SUCCESS
hdfs://192.168.254.128:9000/test/wd01/part-r-00000
hdfs://192.168.254.128:9000/tmp
hdfs://192.168.254.128:9000/tmp/hadoop-yarn/staging
hdfs://192.168.254.128:9000/tmp/hadoop-yarn/staging/history/done_intermediate
hdfs://192.168.254.128:9000/tmp/hadoop-yarn/staging/history/done_intermediate/root/job_1503016403177_00
hdfs://192.168.254.128:9000/tmp/hadoop-yarn/staging/history/done_intermediate/root/job_1503016403177_00
hdfs://192.168.254.128:9000/tmp/hadoop-yarn/staging/history/done_intermediate/root/job_1503016403177_00
hdfs://192.168.254.128:9000/tmp/hadoop-yarn/staging/root/.staging
hdfs://192.168.254.128:9000/tmp/hive/root
hdfs://192.168.254.128:9000/tmp/hive/root/3a6b81a2-026c-464d-ae0b-fae2b70b247e/_tmp_space.db
hdfs://192.168.254.128:9000/tmp/hive/root/5b504b90-eef8-41d8-bcf1-53640bc7f582/_tmp_space.db
hdfs://192.168.254.128:9000/tmp/hive/root/870b0662-ff92-489b-8ef7-e6469596e904/_tmp_space.db
hdfs://192.168.254.128:9000/tmp/hive/root/da639dd1-4b34-4fe5-b0c1-433c0ac887f2/_tmp_space.db
hdfs://192.168.254.128:9000/user
hdfs://192.168.254.128:9000/user/hive/warehouse
```

图 5-24　遍历根目录下的所有文件

5.3.10　根据 filter 获取目录下的文件

HDFS 提供 globStatus 方法对文件路径进行过滤，可以通过使用通配符的方式指定输入，无需列举每个文件和目录来指定输入。Hadoop 为执行通配提供了两个方法：

```
public FileStatus[] globStatus(Path pathPattern) throw IOException
public FileStatus[] globStatus(Path pathPattern, PathFilter filter) throw IOException
```

globStatus()方法返回与路径相匹配的所有文件的 FileStatus 对象数组,并按路径排序。Hadoop 所支持的通配符与 UNIX bash 相同。

第二个方法传入一个 PathFilter 对象作为参数,PathFilter 可以进一步对匹配进行限制。PathFilter 是一个接口,里面只有一个方法 accept(Path path)。具体代码如下:

```java
public class RegexExcludePathFilter implements PathFilter{
    private final String regex;
    public RegexExcludePathFilter(String regex) {
        this.regex = regex;
    }
    @Override
    public boolean accept(Path path) {
        return !path.toString().matches(regex);
    }
}

//通配符的使用
public static void list() throws IOException{
    Configuration conf = new Configuration();
    conf.set("fs.defaultFS", "hdfs://192.168.254.128:9000");
    System.setProperty("HADOOP_USER_NAME", "root");
    FileSystem fs = FileSystem.get(conf);
    //PathFilter 是过滤不符合置顶表达式的路径,下列就是把以 txt 结尾的过滤掉
    FileStatus[] status = fs.globStatus(new Path("/data/*"),
            new RegexExcludePathFilter(".*txt"));
    //FileStatus[] status = fs.globStatus(new Path("hdfs://master:9000/user/hadoop/test/*"));
    Path[] listedPaths = FileUtil.stat2Paths(status);
    for (Path p : listedPaths) {
        System.out.println(p);
    }
}
```

测试代码为:

```java
public static void main(String[] args) throws IllegalArgumentException, Exception {
    list();
}
```

运行结果如图 5-25 所示。

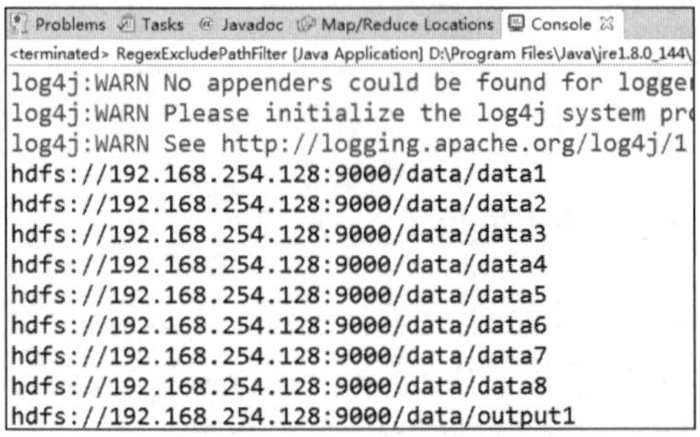

图 5-25 过滤/data 下所有含有.txt 文件的目录

5.3.11 取得数据块所在的位置

通过 FileSystem.getFileBlockLocation(FileStatus file, long start, long len)可以查找指定文件在 HDFS 集群上的位置，其中 file 为文件的完整路径，start 和 len 来标识查找文件的路径。具体实现如下：

```java
/**
 * 取得数据块所在的位置
 * @param path：文件路径
 */
public static void getLocation(Path path) {
    try {
        Configuration conf = new Configuration();
        conf.set("fs.defaultFS", "hdfs://192.168.254.128:9000");
        System.setProperty("HADOOP_USER_NAME", "root");
        FileSystem fs = FileSystem.get(conf);
        FileStatus fileStatus = fs.getFileStatus(path);
        BlockLocation[] blkLocations = fs.getFileBlockLocations(fileStatus, 0, fileStatus.getLen());
        for (BlockLocation currentLocation : blkLocations) {
            String[] hosts = currentLocation.getHosts();
            for (String host : hosts) {
                System.out.println(host);
            }
        }

        //取得最后修改时间
        long modifyTime = fileStatus.getModificationTime();
        Date d = new Date(modifyTime);
        System.out.println(d);
    } catch (Exception e) {
        e.printStackTrace();
    }
}
```

测试代码为：

```java
public static void main(String[] args) throws IllegalArgumentException, Exception {
    getLocation(new Path("/test/log2.txt"));
}
```

运行结果如图 5-26 所示。

```
<terminated> Test1 [Java Application] D:\Program Files\Java\jre1.8.0_144\bin\javaw.exe (2018年1月13日 下午4:39:49)
log4j:WARN No appenders could be found for logger (org.apache.had
log4j:WARN Please initialize the log4j system properly.
log4j:WARN See http://logging.apache.org/log4j/1.2/faq.html#nocor
slave1
slave2
Sat Jan 13 16:26:58 CST 2018
```

图 5-26 文件 log2.txt 在集群节点上的位置

5.4 程序打包

程序调试完毕后，就可以将其打成 JAR 包运行在 Hadoop 集群中，在工程上右击并选择 Export 选项，如图 5-27 和图 5-28 所示。

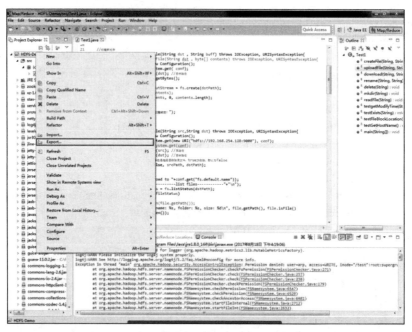

图 5-27　程序打包 1

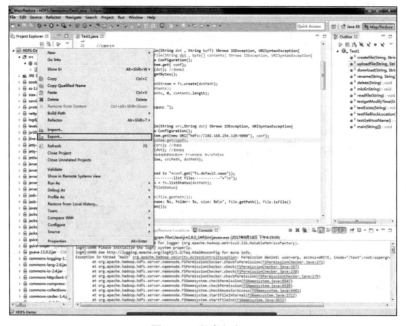

图 5-28　程序打包 2

选择导出的文件路径，如图 5-29 所示。

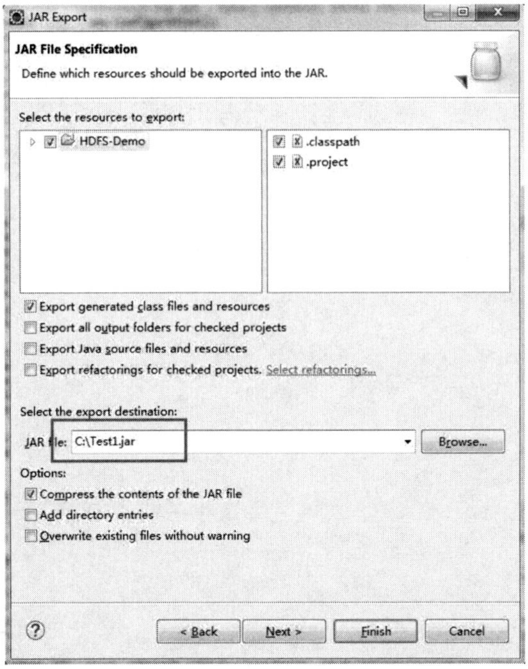

图 5-29　JAR 包输出路径

设置 JAR 包的 Main Class，如果有多个 Main Class，则需要指定唯一一个作为程序的入口，操作如图 5-30 所示。

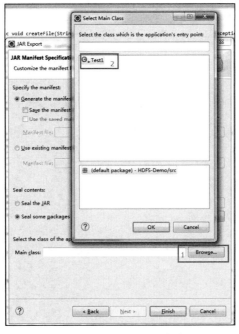

图 5-30　设置 JAR 包主类

其他操作默认即可，最后单击 Finish 按钮，将生成的 Test1.jar 提交到 Hadoop 集群，通过 hadoop jar Test1.jar 即可运行。

5.5 本章小结

HDFS 提供了 Java API 接口，方便用户操作。通常通过以下三步操作即可完成：第一步构造 Configuration 对象并设置相应属性；第二步生成 FileSystem 对象；第三步调用 API 函数。本章讲解了 HDFS Java API 的常用操作，为大家学习相关内容提供了帮助。

第 6 章 并行计算 MapReduce

MapReduce 作为 Hadoop 集群的一个重要组件,是一个能处理和生成超大数据集的算法模型,基于该模型能够容易地编写分布式应用程序,并以一种可靠的、具有容错能力的方式在大量普通配置的计算机上实现并行化处理海量数据集。本章将从 MapReduce 编程模型、工作原理、Yarn 工作流程等方面进行讲解。

6.1 MapReduce 编程模型

6.1.1 并行编程模型概述

并行编程模型是建立在硬件体系结构模型之上的一组程序的抽象,它定义了程序设计与其实现之间的一种形式化的接口,为并行程序的设计者提供了一个简洁而清晰的并行机软硬件系统结构的概念模型。通俗地讲,并行编程模型是指并行编程时程序各模块并行执行时模块间的通信方式。

并行编程模型可以分为两类:进程通信和问题分解,下面简要介绍这两类模型有哪些形式。

1. 进程通信

进程通信涉及并行进程互相通信的机制。最常用的通信形式是共享内存和消息传递,但是通信形式也可以是隐式的,对程序员是不可见的。

(1)共享内存:共享内存是进程间传递数据的一种高效方法。在共享内存模型中,并行进程共享一个进行异步读取的全局地址空间。异步并发访问可能导致条件竞争,因此需要同步机制来避免条件竞争,这些机制包括锁、信号量、管程。传统的多核处理器是直接支持共享内存的,所以导致很多利用该特性的语言和库出现,如 Cilk、OpenMP 和 Threading Building Blocks。

(2)消息传递:在消息传递模型中,并行进程是通过消息传递来交换数据的。这些通信可以是异步的,即消息可以在接收者做好准备前发送,也可以是同步的,即只有接收者准备好接收消息时才能发送。消息传递的 CSP(Communicating Sequential Processes)模型使用同步通信 channel 来连接进程,这种模式被 Occam、Limbo 和 Go 等语言所采用。相反,Actor 模型则使用异步消息传递,这种模式被 D、Scala 和 Salsa 等语言所采用。

(3)隐式通信:在隐式通信中,进程通信对程序员来说是不可见的,进程通信是由编译器或者运行时来处理和实现的。

2. 问题分解

并行程序是由同时运行的进程组成,问题分解涉及所有进程如何被组织起来的方式。问题分解包括三种并行模型:任务并行模型、数据并行模型和隐式并行模型。

(1)任务并行模型:任务并行模型关注进程或线程的执行。这些进程在行为上是不同的,而且相互之间的通信是非常重要的。任务并行模型是表示消息传递通信的一种自然方式,在

Flynn 分类法中，任务并行模型的三种形式是 MIMD（Multiple Instruction Stream Multiple DataStream）、MPMD（Multiple Program Multiple Data）和 MISD（Multiple Instruction Single Data）。

（2）数据并行模型：数据并行模型关注在数据集上执行的操作。一组任务对数据集进行运算，但是会对不同的分区进行运算。在 Flynn 分类法中，任务并行模型的三种形式是 SIMD（Single Instruction Multiple Data）、SPMD（Single Program Multiple Data）和 MIMD。

（3）隐式并行模型：对程序员来说是不可见的，由编译器、运行时或硬件负责实现。例如，在编译器领域，自动并行模型就是将顺序执行的代码转换为并行代码的过程；在计算机体系架构领域，超标量执行就是一种利用指令级并行来实现并行运算的机制。

6.1.2 并行计算编程模型

并行计算或称平行计算是相对于串行计算而言的。它是一种一次可执行多个指令的算法，目的是提高计算速度，以及通过扩大问题求解规模来解决大型而复杂的计算问题。所谓并行计算可以分为时间上的并行和空间上的并行。时间上的并行就是指流水线技术，而空间上的并行则是指用多个处理器并发地执行计算。并行计算是指同时使用多种计算资源解决计算问题的过程，是提高计算机系统计算速度和处理能力的一种有效手段。它的基本思想是用多个处理器来协同求解同一问题，即将被求解的问题分解成若干部分，各部分均由一个独立的处理机来并行计算。并行计算系统既可以是专门设计的、含有多个处理器的超级计算机，也可以是以某种方式互连的若干台独立计算机构成的集群。通过并行计算集群完成数据的处理，再将处理的结果返回给用户。

目前开源社区有许多并行计算模型和框架可供选择，按照实现方式、运行机制、依附的产品生态圈等可以划分为几个类型，每个类型各有优缺点，如果能够对各类型的并行计算框架都进行深入研究及适当的缺点修复，就可以为不同硬件环境下的海量数据分析需求提供不同的软件层面的解决方案。

1. MapReduce

MapReduce 是由谷歌推出的一个编程模型，是一个能处理和生成超大数据集的算法模型，该架构能够在大量普通配置的计算机上实现并行化处理。MapReduce 编程模型结合用户实现的 Map 和 Reduce 函数。用户自定义的 Map 函数处理一个输入的基于 key/value 键值对的集合，MapReduce 把中间所有具有相同 key 值的 value 值集合在一起后传递给 Reduce 函数。用户自定义的 Reduce 函数合并所有具有相同 key 值的 value 值，形成一个较小 value 值的集合。

2. Spark

Spark 是基于 MapReduce 算法实现的分布式计算，拥有 MapReduce 所具有的优点，但不同于 MapReduce 的是中间输出结果可以保存在内存中，从而不再需要读写 HDFS，因此 Spark 能更好地适用于数据挖掘和机器学习等领域。Spark 具有以下特点：

（1）高效性。

1）内存运算。Spark 能够在内存中完成运算，除非特殊要求中间结果均保存在内存中，减少磁盘 IO 操作。根据 Spark 官方提供的数据，如果数据由磁盘读取，速度是 Hadoop MapReduce 的 10 倍以上；如果数据从内存中读取，速度可以高达 100 多倍。

2）Lazy 机制。Spark 运算不是对每一步操作都要计算出结果，而是在执行求和、输出等操作时才会将之前的操作合并执行，在处理大规模数据时能够极大地减少运算量。

3）线程操作。Spark 的作业都是由进程内部的线程池执行，能极大地减少 JVM 开销。

（2）通用性。

1）Spark 既使用 Hadoop HDFS 完成数据存储，也可以使用 HBase 等作为数据源。

2）Spark 能够使用 Yarn 或 Mesos 作为资源管理框架。

3）Spark 不仅支持 Scala 编写应用程序，也支持 Java、Python 和 R 语言进行编写。

（3）可靠性。

Spark 引进了弹性分布式数据集（Resilent Distributed Datasets，RDD），这些集合是弹性的，如果数据集一部分丢失，则可以根据 Lineage（血统）对它们进行重建。

3. Disco

Disco 是一个轻量级的、开源的基于 MapReduce 模型计算的框架。Disco 强大且易于使用，这都要归功于 Python，Disco 分发且复制数据，可以高效安排作业。Disco 甚至拥有能对数以亿计的数据点进行索引以及实时查询的工具。Disco 于 2008 年在 Nokia 研究中心诞生，解决了在大量数据处理方面的挑战。

Disco 支持大数据集的并行计算，在不可靠的计算机集群中，其模式类似于谷歌创造的计算集群。Disco 是一个完美的分析和处理大数据的工具，无需考虑因为分布式带来的技术困难，比如通信协议、负载均衡、锁、作业规划、故障容忍。Disco 可用于多种数据挖掘任务，如大规模分析、建立概率模型、全文索引网页等。

Disco 的核心是用 Erlang 编写的，Erlang 是一种为构建高容错性分布式应用程序而设计的函数语言。Disco 通常使用 Python 作为常用编程语言，这样可以提高开发效率，用更简洁的语法完成更多复杂算法。

4. Phoenix

Phoenix 最早是 Saleforce 的一个开源项目，后来成为 Apache 基金的顶级项目。Phoenix 是构建在 HBase 上的一个 SQL 层，能让人们用标准的 JDBC APIs 而不是 HBase 客户端 APIs 来创建表，插入数据和对 HBase 数据进行查询。Phoenix 完全使用 Java 编写，作为 HBase 内嵌的 JDBC 驱动。Phoenix 查询引擎会将 SQL 查询转换为一个或多个 HBase 扫描，并编排执行以生成标准的 JDBC 结果集。直接使用 HBase API，协同处理器与自定义过滤器，对于简单查询来说，其性能量级是毫秒；对于百万级别的行数来说，其性能量级是秒。Phoenix 可编程多核芯片或共享内存多核处理器（SMPs 和 ccNUMAs），用于数据密集型任务处理。

四种框架优缺点的比较如表 6-1 所示。

表 6-1 四种框架优缺点比较

项目	Hadoop MapReduce	Spark	Phoenix	Disco
编程语言	Java	Scala	C	Erlang
构建平台	需要首先架构基于 Hadoop 的集群系统，通过 HDFS 完成运算的数据存储工作	可以单独运行，也可以与 Hadoop 框架完整结合	独立运行，不需要提前部署集群，运行时系统的实现是建立在 PThread 之上的，也可以方便地移植到其他共享内存线程库上	整个 Disco 平台由分布式存储系统 DDFS 和 MapReduce 框架组成，DDFS 与计算框架高度耦合，通过监控各个节点上的磁盘使用情况进行负载均衡

续表

项目	Hadoop MapReduce	Spark	Phoenix	Disco
功能特点	计算能力非常强，适合超大数据集的应用程序，但是由于系统开销等原因，处理小规模数据的速度不一定比串行程序快，并且本身集群的稳定性不高	在保证容错的前提下，用内存来承载工作集，内存的存取速度快于磁盘多个数量级，从而可以极大地提升性能	利用共享内存缓冲区实现通信，从而避免了因数据复制产生的开销，但Phoenix也存在不能自动执行迭代计算、没有高效的错误发现机制等不足	由一个Master服务器和一系列Worker节点组成，Master和Worker之间采用基于轮询的通信机制，通过HTTP的方式传输数据。轮询的时间间隔不好确定，若时间间隔设置不当，会显著降低程序的执行性能

6.1.3 MapReduce 编程模型

MapReduce 是一种并行计算编程模型，可以使用多种语言开发，例如 Java、Ruby、Python、C++等。MapReduce 通过抽象模型和计算框架把需要做什么（what need to do）与具体怎么做（how to do）分开了，为程序员提供一个抽象和高层的编程接口与框架。程序员仅需要关心其应用层的具体计算问题，仅需编写少量的处理应用本身计算问题的程序代码。如何具体完成这个并行计算任务所相关的诸多系统层细节被隐藏起来，交给计算框架去处理：从分布代码的执行到数千个节点集群的自动调度使用。

（1）map()。map()函数以 key/value 键值对作为输入和输出。MapReduce 框架会自动将输出数据按照 key 的字典升续排序，通常会启动多个 Map 任务执行，Map 任务执行结束后将结果写入本地硬盘，而不是 HDFS 文件系统。

（2）reduce()。reduce()函数以 map()函数的输出结果作为输入，合并相同 key 的 value 值后生成另外一系列 key/value 作为输出并将其写入 HDFS 中。MapReduce 框架会默认启动一个reduce，也可以通过配置设置 0 个或多个 reduce 任务。

MapReduce 提供一个统一的计算框架，可以完成：计算任务的划分和调度、数据的分布存储和划分、处理数据与计算任务的同步、结果数据的收集整理、系统通信、负载平衡、计算性能优化处理和处理系统节点出错检测与失效恢复。

（1）任务调度。提交的一个计算作业（job）将被划分为很多个计算任务（task），任务调度功能主要负责为这些划分后的计算任务分配和调度计算节点（map 节点或 reduce 节点）；同时负责监控这些节点的执行状态，并负责 map 节点执行的同步控制；也负责进行一些计算性能优化处理，如对最慢的计算任务采用多备份执行，选最快完成者作为结果。

（2）数据/代码互定位。为了减少数据通信，一个基本原则是本地化数据处理，即一个计算节点尽可能处理其本地磁盘上所分布存储的数据，这实现了代码向数据的迁移；当无法进行这种本地化数据处理时，再寻找其他可用节点并将数据从网络上传送给该节点（数据向代码迁移），但将尽可能从数据所在的本地机架上寻找可用节点以减少通信延迟。

（3）出错处理。以低端商用服务器构成的大规模 MapReduce 计算集群中，节点硬件（主机、磁盘、内存等）出错和软件有 bug 是常态，因此 MapReduce 需要能检测并隔离出错节点，并调度分配新的节点接管出错节点的计算任务。

（4）分布式数据存储与文件管理。海量数据处理需要一个良好的分布数据存储和文件管理系统支撑，该文件系统能够把海量数据分布存储在各个节点的本地磁盘上，但保持整个数据在逻辑上成为一个完整的数据文件。为了提供数据存储容错机制，该文件系统还要提供数据块的多备份存储管理能力。

（5）Combiner 和 Partitioner。为了减少数据通信开销，中间结果数据进入 reduce 节点前需要进行合并（combiner）处理，把具有同样主键的数据合并到一起避免重复传送。一个 reduce 节点所处理的数据可能会来自多个 map 节点，因此 map 节点输出的中间结果需要使用一定的策略进行适当的划分（partitioner）处理，保证相关数据发送到同一个 reduce 节点。

6.2 MapReduce 工作原理

MapReduce 按照执行的先后顺序，大致分为输入分片、Map、Partitioner、Sort、Combiner、和 Reduce 阶段，其中 Partitioner、Sort 和 Combiner 又通常被称为 Shuffle，工作流程如图 6-1 所示。

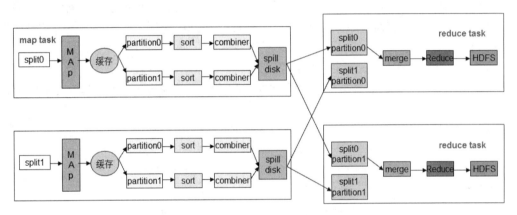

图 6-1　MapReduce 工作流程

（1）输入分片（Input Split）。在进行 Map 计算之前，MapReduce 会根据输入文件计算输入分片，每个输入分片会启动一个 Map 任务，也就是说 Map 任务个数是由输入分片决定的，Map 个数的计算公式为：splitsize=max(minimumsize,min(maximumsize,blocksize))。

blocksize 为数据块大小，minimumsize 和 maximumsize 可以在配置文件中设置，没有设置的话 splitsize 的大小默认等于 blocksize，即 128MB。假设有三个输入文件，大小分别是 64MB、129MB 和 256MB，那么 MapReduce 会把 64MB 的文件划分为一个输入分片，129MB 则是两个输入分片（128MB 和 1MB），256MB 也是两个输入分片（都是 128MB），一共有 5 个输入分片，分别启动 5 个 Map 任务，但是每个 Map 任务运算的数据量不一样。

（2）Map 阶段。通过自定义 map()函数读入数据分片，输出的 key/value 对放到环形内存缓冲区，这个缓冲区专门用来存放 Map 的输出结果，默认大小是 100MB，并且在配置文件里为这个缓冲区设定一个阈值（默认是 0.80，缓冲区大小和阈值都是可配置的）。如果缓冲区的内存达到了阈值的时候，会把内容以一个临时文件的方式存放到磁盘，这个过程叫 spill。另外的 20%内存可以继续写入要写进磁盘的数据，写入磁盘和写入内存操作是互不干扰的，如果缓存区满了，Map 就会阻塞写入内存的操作，让写入磁盘操作完成后再继续执行写入内存操作。

（3）Partitioner 阶段。对 Map 产生的中间结果进行分片，以便将同一分组的数据交给同一个 Reduce 处理。一个 Partitioner 对应一个 Reduce 作业，也就是说有几个 Reduce 就会有几个 Partitioner。Partitioner 用来决定 Map 产生的 key 由哪个 Reduce 来处理，默认处理采用对 key 进行 hash 运算后再对 Reduce 数取模。这种方式只是为了平均 Reduce 的处理能力，但在实际中可能会出现"数据倾斜"的情况。某个 key 有大量相同数据，比如一个 key 有 20W 的数据，而其他所有 key 加起来才有不到 200 条数据，要是由两个 Reduce 处理的话，会造成两个任务处理量严重不均衡，为了避免这种情况就需要用户自定义 Partitioner。

（4）Sort 阶段。根据 key 按照字典升续排序。

（5）Combiner 阶段。Combiner 阶段可以由用户选择是否执行，是一个本地化的 Reduce 操作。它是在 Map 生成中间文件前做一个合并重复 key 值的操作，这样可以减少磁盘存储和网络传输量。不是所有操作都要使用 Combiner，使用原则是 Combiner 的输出不能影响到 Reduce 的最终输入。例如，如果计算只是求总数、最大值或最小值，则可以使用 Combiner，但是求平均值就不能使用 Combiner，否则 Reduce 会求出比真实值大的结果。

（6）Reduce 阶段。需要用户自定义实现，可以没有，比如只是对原始数据做一些格式转换。在 Reduce 执行之前会将从各个 Map 中 Partitioner 相同的数据合并排序，执行的结果保存在 HDFS 上。

下面通过一个单词计数案例来理解各个过程。

（1）读取文件，文件划片后默认会将文件按行分割形成<key,value>，其中 key 的值是行偏移量，如图 6-2 所示。

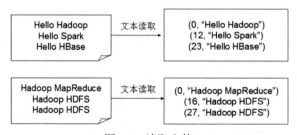

图 6-2　读取文件

（2）执行自定义 map，比如将 1 中的 value 分割成一个个单词，每个单词作为新 key/value 键值对的 key，1 作为 value，如图 6-3 所示。

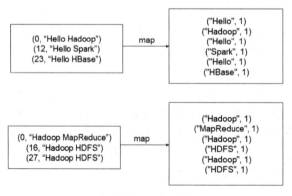

图 6-3　map 操作

（3）得到 map 方法输出的<key,value>对后，Mapper 会执行 shuffle 过程，即先排序后合并，在此默认 reduce 的执行数量为 1，得到 Mapper 的最终输出结果，如图 6-4 所示。

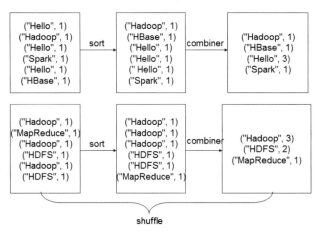

图 6-4　shuffle 操作

（4）Reducer 先对从 Mapper 接收的数据进行合并排序，再交由用户自定义的 Reduce 方法进行处理，得到新的<key,value>对，并将其输出到 HDFS 上，如图 6-5 所示。

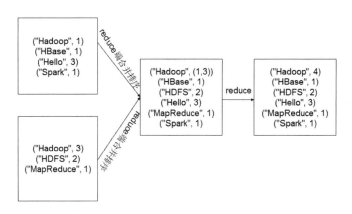

图 6-5　reduce 操作

6.3　Yarn

在 Hadoop 2.x 中采用 Yarn 完成对 MapReduce 框架的资源管理，JobTracker 的工作主要由 ResourceManager 和 ApplicationsManager 代替完成，同时把 Job（作业）的概念换成了 Application（应用程序）。

Yarn 是一个通用的资源调度平台，只负责为运算程序提供资源调度，相当于一个分布式的操作系统平台，不管是 MapReduce 程序还是 Spark、Storm 和 HBase 等都可以在其上运行。

6.3.1　Yarn 基本框架与组件

Yarn 主要由 ResourceManager、NodeManager、Container 和 ApplicationMaster 组件组成。

1. ResourceManager

ResourceManager 是 Yarn 集群的主控节点，负责协调和管理整个集群（所有 NodeManager）的资源，响应用户提交的不同类型应用程序的解析、调度、监控等工作。

ResourceManager 会为每一个 Application 启动一个 ApplicationMaster。它主要由两个组件构成：调度器（Scheduler）和应用程序管理器（ApplicationsManager，ASM）。

（1）Scheduler。

调度器根据应用程序的资源需求进行资源分配，不参与应用程序具体的执行和监控等工作。调度器是一个可插拔的组件，用户可以根据自己的需求实现自己的调度器。Yarn 本身提供了多种直接可用的调度器，比如 FIFO、Fair Scheduler 和 Capacity Scheduler 等。

（2）ApplicationsManager。

应用程序管理器 ASM 负责管理整个 Yarn 中所有的应用程序，包括应用程序提交、与调度器协商资源以启动 ApplicationMaster、监控 ApplicationMaster 运行状态并在失败时重新启动它等。

2. NodeManager

NodeManager 是 Yarn 集群的从节点。它是 Yarn 集群当中资源的提供者，也是真正执行应用程序的容器（Container）的提供者。NodeManager 负责接收 ResourceManager 的资源分配请求，监控应用程序的资源使用情况（CPU、内存、硬盘、网络），并通过心跳向集群资源调度器 ResourceManager 汇报 Container 的使用信息，通过与 ResourceManager 配合，负责整个 Hadoop 集群中的资源分配工作。

3. ApplicationMaster

ApplicationMaster 是应用级别的，一个 ApplicationMaster 对应一个应用程序。它主要负责向 ResourceManager 申请执行任务的资源容器，运行任务，监控整个任务的执行，跟踪整个任务的状态，处理任务失败等异常情况。

6.3.2 Yarn 工作流程

如图 6-6 所示是 Yarn 的工程流程。

（1）客户端通过 ResourceManager 向 ApplicationsManager 提交应用，将 JAR 包文件等上传到 HDFS 的指定目录，并请求一个 ApplicationMaster 实例。

（2）ApplicationsManager 生成一个 Application ID。ResourceManager 将请求转发给调度器，调度器分配一个 Container，然后 ResourceManager 在这个 Container 中启动 ApplicationMaster 实例，并交由 NodeManager 对 ApplicationMaster 实例进行管理。

（3）ApplicationMaster 向 ResourceManager 进行注册，注册之后客户端就可以通过查询 ResourceManager 获得 ApplicationMaster 的详细信息。

（4）ApplicationMaster 通过计算任务数和数据本地性等信息向 ResourceManager 申请 Container，并由 ApplicationMaster 负责将 Container 启动。

（5）应用程序的代码在启动的 Container 中运行，NodeManager 把运行的进度、状态等信息发送给 ApplicationMaster。

（6）重复步骤（4）和（5），直至应用程序运行结束。ApplicationMaster 向 ResourceManager 取消注册然后关闭，所有用到的 Container 也归还给系统。

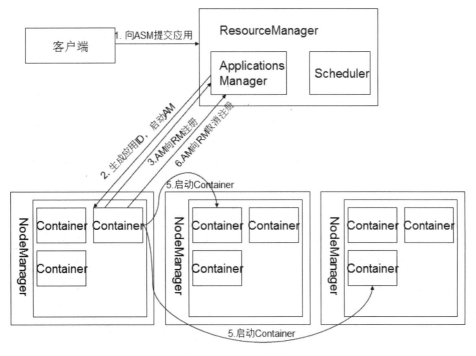

图 6-6　Yarn 工作流程

Hadoop 2.x MapReduce 框架采用 Yarn 作为资源管理,在运行时有它特有的两类实例进程:MRAppMaster 和 Yarnchild,其工作流程和 Yarn 很相似。在 MapReduce 中 ApplicationMaster 的实例化叫做 MRAppMaster,MRAppMaster 负责整个程序的过程调度及状态协调。MRAppMaster 启动后根据本次应用的描述信息计算出需要的 Map 和 Reduce 任务数量,然后向集群申请机器启动相应数量的进程,这个进程称为 Yarnchild。MRAppMaster 首先启动 Map 任务的 Yarnchild,当监控到所有 Map Yarnchild 任务完成之后再去启动 Reduce 任务的 Yarnchild 并释放 Map 任务使用资源(真实情况是,某些 Map Yarnchild 处理完成后,就会开始启动 Reduce Yarnchild,可以观察 3.3 节 PI 的运行实例即为该种情况)。当 Reduce Yarnchild 运行完毕后,MRAppMaster 和 Yarnchild 都会被撤销并释放相应资源。

6.3.3　新旧 Hadoop MapReduce 框架对比

通过对新旧 MapReduce 框架做详细的分析和对比,可以看到有下述几点显著变化。

客户端不变,其调用 API 及接口大部分保持兼容,这也是为了对开发使用者透明化,使其不必对原有代码做大的改变,但是原框架中核心的 JobTracker 和 TaskTracker 不见了,取而代之的是 ResourceManager、ApplicationMaster 和 NodeManager 三个部分。

Yarn 框架相对于老的 MapReduce 框架有什么优势呢?

这个设计大大减小了 JobTracker(也就是现在的 ResourceManager)的资源消耗,并且让监测每一个 Job 子任务(task)状态的程序分布式化了,更安全、更优美。

在新的 Yarn 中,ApplicationMaster 是一个可变更的部分,用户可以对不同的编程模型写自己的 AppMaster,让更多类型的编程模型能够跑在 Hadoop 集群中。对于资源的表示以内存为单位(在目前版本的 Yarn 中,没有考虑 CPU 的占用),比之前以剩余 slot 的数目更合理。

旧的框架中，JobTracker 一个很大的负担就是监控 Job 下的 tasks 的运行状况，现在这个部分就扔给 ApplicationMaster 做了，而 ResourceManager 中有一个模块叫做 ApplicationsMaster（注意不是 ApplicationMaster），它是监测 ApplicationMaster 的运行状况，如果出问题，会将其在其他机器上重启。

Container 是 Yarn 为了将来作资源隔离而提出的一个框架。这一点应该是借鉴了 Mesos 的工作，目前是一个框架，仅仅提供 Java 虚拟机内存的隔离，既然资源表示成内存量，那就没有了之前的 map slot/reduce slot 分开造成集群资源闲置的尴尬情况。

新的 Yarn 框架相对旧的 MapRduce 框架而言，其配置文件、启停脚本及全局变量等也发生了一些变化，主要的改变如表 6-2 所示。

表 6-2 新旧 Hadoop 脚本/变量/位置变化表

改变项	原框架中	新框架中（Yarn）
配置文件位置	${hadoop_home_dir}/conf	${hadoop_home_dir}/etc/hadoop/
启停脚本	${hadoop_home_dir}/bin/start(stop)-all.sh	${hadoop_home_dir}/sbin/start(stop)-dfs.sh ${hadoop_home_dir}/bin/start(stop)-all.sh
JAVA_HOME 全局变量	${hadoop_home_dir}/bin/start-all.sh 中	${hadoop_home_dir}/etc/hadoop/hadoop-env.sh ${hadoop_home_dir}/etc/hadoop/Yarn-env.sh
HADOOP_LOG_DIR 全局变量	不需要配置	${hadoop_home_dir}/etc/hadoop/hadoop-env.sh

新的 Yarn 框架与原 Hadoop MapReduce 框架相比变化较大，核心的配置文件中很多项在新框架中已经废弃，而新框架中新增了很多其他配置项，看表 6-3 会更加清晰。

表 6-3 新旧 Hadoop 框架配置项变化表

配置文件	配置项	原框架中	新框架中（Yarn）
core-site.xml	系统默认分布式文件 URI	fs.default.name	fs.defaultFS
hdfs-site.xml	HDFS NameNode 存放 name table 的目录	dfs.name.dir	dfs.namenode.name.dir
	HDFS DataNode 存放数据 block 的目录	dfs.data.dir	dfs.datanode.data.dir
	分布式文件系统数据块复制数	dfs.replication	dfs.replication
mapred-site.xml	Job 监控地址及端口	mapred.job.tracker	无
	第三方 MapReduce 框架	无	mapreduce.framework.name
Yarn-site.xml	ResourceManager Address	无	Yarn.resourcemanager.address
	Scheduler Address	无	Yarn.resourcemanager.scheduler.address
	ResourceManager WEB address	无	Yarn.resourcemanager.webapp.address
	Resource Tracker Address	无	Yarn.resourcemanager.resource-tracker.address

6.4 MapReduce Shuffle 性能调优

1. Map 端优化

（1）io.sort.mb。

一个 Map 都会对应存在一个内存缓冲区 kvbuffer，kvbuffer 默认为 100MB，大小可以根据 Job 提交时设定的参数 io.sort.mb 来调整。当 Map 产生的数据非常大，并且把 io.sort.mb 调大，那么在 Map 计算过程中 Spill 的次数就会降低，Map task 对磁盘的操作就会变少，如果 Map task 的瓶颈在磁盘上，这样调整就会大大提高 Map 的计算性能。

（2）io.sort.spill.percent。

io.sort.spill.percent 控制的是 kvbuffer 开始 spill 到磁盘的阈值，默认为 0.80。这个参数同样也是影响 spill 频繁程度，进而影响 Map task 运行周期对磁盘的读写频率。

（3）io.sort.factor。

Merge 的过程中，有一个参数 io.sort.factor 可以调整这个过程的行为，默认为 10。该参数表示当 merge spill 文件时，最多能有多少并行的 stream 向 merge 文件中写入。如果 Map 的中间结果非常大，调大 io.sort.factor，有利于减少 merge 次数，进而降低 Map 对磁盘的读写频率，有可能达到优化作业的目的。

（4）min.num.spill.for.combine。

当 job 指定了 Combiner 时，会在 Map 端根据 Combiner 定义的函数将 Map 结果进行合并。运行 Combiner 函数的时机有可能会是 merge 完成之前或之后。这个时机由 min.num.spill.for.combine 参数控制，默认为 3。通过这样的方式，就可以在 spill 非常多需要 merge，并且很多数据需要做 Combine 的时候，减少写入到磁盘文件的数据数量，同样降低了对磁盘的读写频率，达到优化作业的目的。

（5）mapred.compress.map.output。

将这个参数设置为 true 时，那么 Map 在写中间结果时就会先将数据压缩后再写入磁盘，读结果时也会先解压后读取数据。这样做的结果就是：写入磁盘的中间结果数据量会变少，但是 CPU 会消耗一些用来压缩和解压。所以，这种方式通常适合 Job 中间结果非常大，瓶颈不在 CPU，而是在磁盘读写的情况。

（6）mapred.map.output.compression.codec。

当采用 Map 中间结果压缩的情况下，通过 mapred.map.output.compression.codec 参数可以选择不同的压缩模式，现有支持的压缩格式有 GzipCodec、LzoCodec、BZip2Codec、LzmaCodec 等。通常来说，想要达到比较平衡的 CPU 和磁盘压缩比，LzoCodec 比较适合，但也要取决于 Job 的具体情况。

2. Reduce 端优化

（1）mapred.reduce.parallel.copies。

对一个 Reduce 来说，可以从并行的多个 Map 中下载，并行度通过参数 mapred.reduce.parallel.copies 进行设置。默认为 5 个并行的下载线程，这个参数比较适合 Map 很多并且完成得比较快的 Job 的情况下，有利于 Reduce 更快地获取属于自己部分的数据。

（2）mapred.reduce.copy.backoff。

Reduce 下载线程的最大下载时间段通过参数 mapred.reduce.copy.backoff 设置，默认为 300 秒。如果超过该时间，Reduce 下载线程中断，并尝试从其他地方下载。如果集群环境的网络本身是瓶颈，那么用户可以通过调大这个参数来避免 Reduce 下载线程被误判为失败的情况。不过在网络环境比较好的情况下，没有必要调整。

（3）io.sort.factor。

Reduce 将 Map 结果下载到本地时同样也是需要进行 merge，所以 io.sort.factor 的配置选项同样会影响 Reduce 进行 merge 时的行为，可能通过调大这个参数来加大一次 merge 时的并发吞吐量，优化 Reduce 效率。

（4）mapred.job.shuffle.merge.percent。

mapred.job.shuffle.merge.percent 这个限度阈值可以控制 merge 的开始，默认为 0.66。如果下载速度很快，很容易就把内存缓存撑大，那么调整一下这个参数有可能会对 Reduce 的性能有所帮助。

（5）mapred.job.reduce.input.buffer.percent。

mapred.job.reduce.input.buffer.percent 参数默认为 0，表示 Reduce 是全部从磁盘开始读并处理数据。如果这个参数大于 0，那么就会有一定量的数据被缓存在内存并输送给 Reduce，当 Reduce 计算逻辑消耗内存很小时，可以分一部分内存用来缓存数据。

6.5 本章小结

本章介绍了并行计算 MapReduce 的相关知识。MapReduce 是将复杂的、运行于大规模集群上的并行计算过程简化为 Map 和 Reduce。MapReduce 的执行过程主要分为以下几个阶段：从 HDFS 读入数据，执行 Map 任务，分区、排序、合并（这三个阶段统称为 Shuffle），执行 Reduce 任务并将最终结果输出。Hadoop 2.0 采用 Yarn 作为资源管理框架，只负责为运算程序提供资源调度，简化了 MapReduce 的任务流程。在本章最后针对 Shuffle 性能调优提出了一些优化建议。

第 7 章 MapReduce Java API 编程

MapReduce 同样对外提供的有 Java API 接口以方便用户开发,MapReduce 主要有 6 个可编程组件,分别是 InputFormat、Mapper、Combiner、Partitioner、Reducer 和 OutputFormat,很多时候用户只需要编写 Map 和 Reduce 就可以快速完成并行化程序设计,但是对于复杂一点的程序就需要自定义编程组件来完成。本章将分别介绍 MapReduce Java API 接口、应用实例和高级编程。

7.1 MapReduce Java API 接口讲解

在编写 MapReduce 程序时,用户分别通过 InputFormat 和 OutputFormat 指定输入和输出格式,并定义 Mapper 和 Reducer 指定 Map 阶段和 Reduce 阶段要做的工作。在 Mapper 或者 Reducer 中,用户只需指定一对 key/value 的处理逻辑,Hadoop 框架会自动顺序迭代解析所有 key/value,并将每对 key/value 交给 Mapper 或者 Reducer 处理。表面上来看,Hadoop 限定数据格式必须为 key/value 形式,过于简单,很难解决复杂问题,实际上,可以通过组合的方法使 key 或者 value(比如在 key 或者 value 中保存多个字段,每个字段用分隔符分开,或者 value 是个序列化后的对象,在 Mapper 中使用时将其反序列化等)保存多重信息,以解决输入格式较复杂的应用。

MapReduce 编程模型对外提供的编程接口体系结构如图 7-1 所示,整个编程模型位于应用程序层和 MapReduce 执行器之间,可以分为两层:第一层是最基本的 Java API,主要有 6 个可编程组件,分别是 InputFormat、Mapper、Combiner、Partitioner、Reducer 和 OutputFormat,Hadoop 自带了很多直接可用的 InputFormat、Partitioner 和 OutputFormat,大部分情况下用户只需编写 Mapper 和 Reducer 即可;第二层是工具层,位于基本 Java API 之上,主要是为了方便用户编写复杂的 MapReduce 程序和利用其他编程语言增加 MapReduce 计算平台的兼容性而提出来的,在该层中,主要提供了 4 个编程工具包。

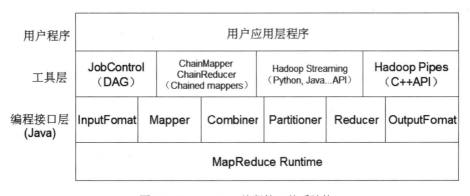

图 7-1 MapReduce 编程接口体系结构

（1）JobControl：方便用户编写有依赖关系的作业，这些作业往往构成一个有向图，所以通常称为 DAG（Directed Acyclic Graph）作业。

（2）ChainMapper/ChainReducer：方便用户编写链式作业，即在 Map 或者 Reduce 阶段存在多个 Mapper，形式如下：

[MAPPER+ REDUCER MAPPER*]

（3）Hadoop Streaming：方便用户采用非 Java 语言编写作业，允许用户指定可执行文件或者脚本作为 Mapper/Reducer。

（4）Hadoop Pipes：专门为 C/C++程序员编写 MapReduce 程序提供的工具包。

7.1.1 InputFormat 接口

InputFormat 是 MapReduce 中一个很常用的概念，它在程序的运行中到底起到了什么作用呢？

InputFormat 接口决定了输入文件如何被 Hadoop 分块与接收。InputFormat 能够从一个 Job 中得到一个 split 集合（InputSplit[]），然后再为这个 split 集合配上一个合适的 RecordReader（getRecordReader）来读取每个 split 中的数据。InputFormat 其实是一个接口，包含了两个方法：

```
public interface InputFormat<K, V> {
    InputSplit[] getSplits(JobConf job, int numSplits) throws IOException,Interrupted Exception;
    RecordReader<K, V> createRecordReader(InputSplit split,
        TaskAttemptContext context) throws IOException,InterruptedException;
}
```

（1）getSplits(JobContext context)方法：负责将一个大数据逻辑分成许多片。比如数据库表有 100 条数据，按照主键 ID 升序存储。假设每 20 条分成一片，这个 List 的大小就是 5，然后每个 InputSplit 记录两个参数，第一个为这个分片的起始 ID，第二个为这个分片数据的大小，这里是 20。很明显 InputSplit 并没有真正存储数据，只是提供了一个如何将数据分片的方法。

（2）createRecordReader(InputSplit split,TaskAttemptContext context)方法：根据 InputSplit 定义的方法返回一个能够读取分片记录的 RecordReader。getSplit 用来获取由输入文件计算出来的 InputSplit，后面会看到计算 InputSplit 时会考虑输入文件是否可分割、文件存储时分块的大小和文件大小等因素。而 createRecordReader()提供了前面说的 RecordReader 的实现，将 key/value 对从 InputSplit 中正确读出来，比如 LineRecordReader，它是以偏移值为 key，每行的数据为 value，这使所有 createRecordReader()返回 LineRecordReader 的 InputFormat 都是以偏移值为 key，每行数据为 value 的形式读取输入分片的。

其实，很多时候并不需要我们实现 InputFormat 来读取数据，Hadoop 自带有很多数据输入格式，已经实现了 InputFormat 接口。

1．InputFormat 接口实现类

InputFormat 接口实现类有很多，主要有以下几个：FileInputFormat、TextInputFormat、KeyValueTextInputFormat、NlineInputFormat 和 SequenceFileInputFormat。

（1）FileInputFormat。

FileInputFormat 是所有使用文件作为其数据源的 InputFormat 实现的基类，它的主要作用

是指出作业的输入文件位置（虽然在 Hadoop 2.0 中作业已经改为程序，但是为了统一称呼，这里仍称为作业）。因为作业的输入被设定为一组路径，这对指定作业输入提供了很强的灵活性。FileInputFormat 提供了 4 种静态方法来设定 Job 的输入路径：

  public static void addInputPath(Job job,Path path);
  public static void addInputPaths(Job job,String commaSeparatedPaths);
  public static void setInputPaths(Job job,Path... inputPaths);
  public static void setInputPaths(Job job,String commaSeparatedPaths);

addInputPath()方法可以将一个或多个路径加入路径列表，可以分别调用这两种方法来建立路径列表；setInputPaths()方法一次设定完整的路径列表，替换前面调用中在 Job 上所设置的所有路径。它们具体的使用方法如下：

  //设置一个源路径
  FileInputFormat.addInputPath(job, new Path("/inputPath1"));
  //设置多个源路径，多个源路径之间用逗号分开
  FileInputFormat.addInputPaths(job, "/inputPath1, /inputPath2,...");
  // inputPaths 是一个 Path 类型的数组，可以包含多个源路径，比如/inputPath1、/inputPath2 等
  FileInputFormat.setInputPaths(job, inputPaths);
  //设置多个源路径，多个源路径之间用逗号分开
  FileInputFormat.setInputPaths(job, "/inputPath1, /inputPath2,...");

add 方法、set 方法允许指定包含的文件。如果需要排除特定文件，可以使用 FileInputFormat 的 setInputPathFilter()方法设置一个过滤器。

  public static void setInputPathFilter(Job job,Class<? extends PathFilter filter);

即使不设置过滤器，FileInputFormat 也会使用一个默认的过滤器来排除隐藏文件。如果通过调用 setInputPathFilter()设置了过滤器，它会在默认过滤器的基础上进行过滤。换句话说，自定义的过滤器只能看到非隐藏文件。

对于输入的数据源是文件类型的情况下，Hadoop 不仅擅长处理非结构化文本数据，而且可以处理二进制格式的数据，但它们的基类都是 FileInputFormat。下面介绍的几种常用输入格式都实现了 FileInputFormat 基类。

（2）TextInputFormat。

TextInputFormat 是默认的 InputFormat。每条记录是一行输入，键是 LongWritable 类型，存储该行在整个文件中的字节偏移量，值是这行的内容，不包括任何行终止符（换行符、回车符），它被打包成一个 Text 对象。

比如，一个分片包含了如下 5 条文本记录，记录之间使用 "/t"（水平制表符）分隔。

  1 1
  2 2
  3 3
  4 4
  5 5

每条记录表示为以下键/值对：

  (0, 1 1)
  (3, 2 2)
  (6, 3 3)
  (9, 4 4)
  (12, 5 5)

很明显,键并不是行号。一般情况下,很难取得行号,因为文件按字节而不是按行切分为分片。

(3) KeyValueTextInputFormat。

每一行均为一条记录,被分隔符(默认是(\t))分隔为 key(Text)和 value(Text)。可以通过 mapreduce.input.keyvaluelinerecordreader.key.value.separator 属性来设定分隔符。

比如,一个分片包含了如下 5 条文本记录,记录之间使用"/t"分隔。

1   1
2   2
3   3
4   4
5   5

每条记录表示为以下键/值对:

(1, 1)
(2, 2)
(3, 3)
(4, 4)
(5, 5)

此时的键是每行排在制表符之前的 Text 序列。

(4) NLineInputFormat。

通过 TextInputFormat 和 KeyValueTextInputFormat,每个 Mapper 收到的输入行数不同。行数取决于输入分片的大小和行的长度。如果希望 Mapper 收到固定行数的输入,需要将 NLineInputFormat 作为 InputFormat。与 TextInputFormat 一样,键是文件中行的字节偏移量,值是行本身,N 是每个 Mapper 收到的输入行数。N 设置为 1(默认值)时,每个 Mapper 正好收到一行输入。mapreduce.input.lineinputformat.linespermap 属性实现 N 值的设定。

下面是一个示例,仍然以上面的 5 行输入为例。

1   1
2   2
3   3
4   4
5   5

例如,如果 N 是 3,则每个输入分片包含 3 行。一个 Mapper 收到 3 行键值对:

1   1
2   2
3   3

另一个 mapper 则收到后两行(因为总共才 5 行,所有另一个 Mapper 只能收到两行):

4   4
5   5

这里的键和值与 TextInputFormat 生成的一样。

(5) SequenceFileInputFormat。

用于读取 Sequence File,键和值由用户定义。序列文件为 Hadoop 专用的压缩二进制文件格式,它专用于在一个 MapReduce 作业和其他 MapReduce 作业之间传送数据(适用于多个 MapReduce 链接操作)。

2. DBInputFormat

这种输入格式用于使用 JDBC 从关系数据库中读取数据。因为它没有任何共享能力，所以在访问数据库的时候必须非常小心，在数据库中运行太多的 Mapper 读数据可能会使数据库受不了。正是由于这个原因，DBInputFormat 最好用于加载少量的数据集。与之相对应的输出格式是 DBOutputFormat，它适用于将作业输出数据（中等规模的数据）转存到数据库。

3. 自定义 InputFormat

有时候 Hadoop 自带的输入格式并不能完全满足业务的需求，所以需要我们根据实际情况自定义 InputFormat 类。而数据源一般都是文件数据，那么自定义 InputFormat 时继承 FileInputFormat 类会更为方便，从而不必考虑如何分片等复杂操作。自定义输入格式分为以下几步：

（1）继承 FileInputFormat 基类。
（2）重写 FileInputFormat 里面的 isSplitable()方法。
（3）重写 FileInputFormat 里面的 createRecordReader()方法。

上述步骤如何自定义输入格式呢？下面我们通过一个示例来加强理解。

取有一份学生三门课程的期末考试成绩数据，现在我们希望统计每个学生的总成绩和平均成绩。样本数据如下，每行数据的数据格式为：学号、姓名、语文成绩、英语成绩、数据库成绩。

```
1615925362    Lucy     76    80    88
1615925363    John     87    77    98
1615925364    James    65    75    74
1615925365    Lily     87    88    86
1615925366    Lilei    76    77    75
1615925367    Smith    78    68    80
```

下面我们就编写程序，实现自定义输入并求出每个学生的总成绩和平均成绩。分为以下几个步骤（这里只给出步骤）：

第一步：为了便于每个学生学习成绩的计算，这里我们需要自定义一个 ScoreWritable 类实现 WritableComparable 接口，将学生各门成绩封装起来。

第二步：自定义输入格式 ScoreInputFormat 类，首先继承 FileInputFormat，然后分别重写 isSplitable()方法和 createRecordReader()方法。需要注意的是，重写 createRecordReader()方法其实也就是重写其返回的对象 ScoreRecordReader。ScoreRecordReader 类继承 RecordReader，实现数据的读取。

第三步：编写 MapReduce 程序，统计学生总成绩和平均成绩。需要注意的是，上面我们自定义的输入格式 ScoreInputFormat 需要在 MapReduce 程序中作如下设置：job.setInputFormatClass(ScoreInputFormat.class)，设置自定义输入格式。

一般情况下，并不需要我们自定义输入格式，Hadoop 自带有很多种输入格式，基本满足工作的需要。

7.1.2 Mapper 类

Mapper 类就是实现 Map 任务的类。Hadoop 提供了一个抽象的 Mapper 基类，程序员需要继承这个基类并实现其中相关的接口函数。

一个示例 Mapper 类的定义如下：

　　public static class MyMapper extend Mapper<Object, Text ,Text ,IntWritable>

（1）Mapper 类是 Hadoop 提供的一个抽象类，程序员可以继承这个基类并实现其中的相关接口函数。它位于 org.apache.hadoop.mapreduce.Reducer<KEYIN, VALUEIN, KEYOUT, VALUEOUT>，在 Mapper 中实现的是对大量数据记录或元素的重复处理，并对每个记录或元素做感兴趣的处理或取感兴趣的中间结果。

（2）Mapper 类中的 4 个方法。

　　protected void setup(Context context)
　　protexted void map(KEYIN key, VALUEIN value, Context context)
　　protected void cleanup(Context context)
　　public void run(Context context)

其中 setup()方法一般是用于 Mapper 类实例化时用户程序可能需要做的一些初始化工作（如创建一个全局数据结构、打开一个全局文件、建立数据库连接等）；map()方法一般承担主要的处理工作；cleanup()方法是收尾工作，如关闭文件或者执行 map()后的键值对的分发等。

（3）map()方法。

　　public void map(Object key , Text value , Context context) throws IOException,InterruptedException

其中 key 是传入 map 的键值，value 是对应键值的 value 值，context 是环境对象参数，供程序访问 Hadoop 的环境对象。

map()方法对输入的键值对进行处理，产生一系列的中间键值对，转换后的中间键值对可以有新的键值类型。输入的键值对可以根据实际应用设定，例如文档数据记录可以是文本文件中的行或数据表格中的行。

Hadoop 使用 MapReduce 框架为每个由作业 InputFormat 产生的 InputSplit 生成一个 Map 任务，Mapper 类可以通过 JobContext.getConfiguration()访问作业的配置信息。

```
public static class TokenizeMapper extends Mapper<Object , Text, Text, IntWritable>{
    private final static IntWritable one = new IntWritable(1);
    private Text word = new Text();
    //完成词频统计的 map 方法
    public void map() throws IOException ,InterruptedException{
        StringTokenizer itr = new   StringTokenizer(value.toString());
        while(){
            word.set(itr.nextToken());
            context.write(word,one);
        }
    }
}
```

（4）setup()和 cleanup()。

Mapper 类在实例化时将调用一次 setup()方法做一些初始化 Mapper 类的工作，例如程序需要时可以在 setup()方法中读入一个全局参数、装入一个文件、连接一个数据库。然后系统会为 InputSplit 中的每一个键值对调用 map()方法，执行程序员编写的计算逻辑。最后，系统将调用一次 cleanup()方法为 Mapper 类做一些结束清理工作，如关闭在 setup()中打开的文件或建立的数据库连接。而默认情况下，这两个函数什么都不做，除非用户重载实现。

注意：setup()和 cleanup()仅仅在初始化 Mapper 实例和 Mapper 任务结束时由系统作为回调函数分别各做一次，而不是每次调用 map()方法时都去执行一次。

（5）Mapper 输出结果的整理。由一个 Mapper 节点输出的键值对首先会进行合并处理，以便 key 相同的键值对合并为一个键值对，这样做的目的是为了减少大量键值对在网络上传输的开销。系统提供了一个 Combiner 类来完成这个合并过程，用户还可以定制并制定一个自定义的 Combiner，通过 JobConf.setCombinerClass(Class)来设置具体所使用的 Combiner 对象。

（6）Mapper 输出的中间键值对还需要进行一些整理，以便将中间结果键值对传递给 Reduce 节点进行后续处理，这个过程也称为 Shuffle。这个整理过程中会将 key 相同的 value 构成的所有键值对分到同一组。Haddop 提供了一个 Partitioner 类来完成这个分组处理过程。用户可以通过实现一个自定义的 Partitioner 来控制哪些键值对发送到对应 Reduce 节点。

（7）在传送给 Reduce 节点之前，中间结果键值对还需要按照 key 值进行排序，以便于后续的处理。这个排序过程将由一个 Sort 类来完成，用户可以通过 JobConf.setOutputKeyConparatorClass(Class)来指定 Sort 类的比较器，从而控制排序的顺序，但如果是使用默认的比较器，则不需要进行这个设置。

（8）Shuffle 之后的结果会被分给各个 Reduce 节点。简单地说，Combiner 是为了减少数据通信开销，中间结果数据进入 Reduce 节点前进行合并处理，把具有同样主键的数据合并到一起避免重复传送。此外，一个 Reduce 节点所处理的数据可能会来自多个 Map 节点，因此，Map 节点输出的中间结果需要使用一定的策略进行适当的分区（Partitioner）处理，保证相关数据发送到同一个 Reduce 节点。

（9）实际上 Combiner 类是在 Map 节点上执行的，而 Partitioner 和 Sort 是在 Reduce 节点上执行的。

7.1.3 Partitioner 类

Partitioner 的作用是对 Mapper 产生的中间结果进行分片，以便将同一分组的数据交给同一个 Reduce 处理，它直接影响 Reduce 阶段的负载均衡。用户需要继承该接口实现自己的 Partitioner 以指定 Map task 产生的 key/value 对交给哪个 Reduce task 处理，好的 Partitioner 能让每个 Reduce task 处理的数据相近，从而达到负载均衡。Partitioner 接口定义如下：

```
public abstract class Partitioner<KEY, VALUE> {

    /**
     * Get the partition number for a given key (hence record) given the total
     * number of partitions i.e. number of reduce-tasks for the job.
     * 通过给定总的分区数（即一般为 Reduce 的个数）获得每个关键字 Key 所对应的分区（所对应的 Reduce 上）。
     * <p>Typically a hash function on a all or a subset of the key.</p>
     *
     * @param key the key to be partioned. 关键字
     * @param value the entry value.
     * @param numPartitions the total number of partitions. 一般是 Reduce 的个数
     * @return the partition number for the <code>key</code>. 哪个 Reduce
```

```
        */
        public abstract int getPartition(KEY key, VALUE value, int numPartitions);
    }
```

MapReduce 提供了两个 Partitioner 实现:HashPartitioner 和 TotalOrderPartitioner。其中 HashPartitioner 是默认实现,它实现了一种基于哈希值的分片方法,代码如下:

```
        public int getPartition(K2 key, V2 value, int numReduceTasks) {
            return (key.hashCode() & Integer.MAX_VALUE) % numReduceTasks;
        }
```

TotalOrderPartitioner 提供了一种基于区间的分片方法,通常用在数据全排序中。

7.1.4 Combiner 类

MapReduce 框架使用 Mapper 将数据处理成一个个<key, value>键值对,再对其进行合并和处理,最后使用 Reduce 处理数据并输出结果。上述过程会受限于集群的带宽,会有网络瓶颈,比如,在做词频统计的时候,大量具有相同主键的键值对数据如果直接传送到 Reduce 节点会引起较大的网络带宽开销。可以对每个 Map 节点处理完成的中间键值对做一个压缩,即把那些主键相同的键值对归并为该键名下的一组数值,这样做不仅可以减轻网络压力,同样也可以大幅提高程序的效率。

MapReduce 通常在 Mapper 类结束后传入 Reduce 节点之前使用一个 Combiner 类来解决相同主键键值对的合并处理。Combiner 的作用主要是为了合并和减少 Mapper 的输出从而减少网络带宽和 Reduce 节点上的负载。如果我们定义一个 Combiner 类,MapReduce 框架会使用它对中间数据进行多次的处理。

Combiner 类在实现上类似于 Reducer 类,事实上它就是一个与 Reducer 类一样,继承自 Reducer 基类的子类。Combiner 的作用只是为了解决网络通信性能问题,因此使用不使用 Combiner 对结果应该是没有任何影响的。为此,需要特别注意的是,程序设计时,为了保证使用 Combiner 后完全不影响 Reducer 的处理和最终结果,Combiner 不能改变 Mapper 类输出的中间键值对的数据类型。

如果 Reducer 只运行简单的分布式的聚集方法,例如最大值、最小值或者计数,由于这些运算与 Combiner 类要做的事情是完全一样的,因此在这种情况下可以直接使用 Reducer 类作为 Combiner 使用,否则将会出现完全错误的结果。在这种情况下,需要定制一个专门的 Combiner 类来完成合并处理。

假设第一个 map 输出如下:
    (2017, 1)
    (2017, 2)
    (2018, 3)
第二个 map 输出如下:
    (2017, 1)
    (2018, 2)
reduce 函数被调用时,其输入是:
    (2017, [1, 2, 1])
    (2018, [3, 2])

结果为：

(2017, 2)

(2018, 3)

如果调用 combine 函数像 reduce 函数一样去寻找每个 map 输出的最大值，那么输出结果应该是：

(2017, [2])

(2018, [3])

reduce 输出结果和以前一样。可以通过下面的表达式来说明最大值的函数调用：

max(1, 2, 1) = max(max(1, 2), max(1)) = max(2, 1) = 2

max(3,2) = max(max(3), max(2)) = 3

并不是所有函数都有这个属性。例如，我们计算平均值就不能使用平均函数作为 combiner。

mid(1, 2, 1) = 1.33

mid(3, 2) = 2.5

但是：

mid(mid(1, 2), mid(1)) = 1.25

mid(mid(3), mid(2)) = 2.5

Combiner 函数不能取代 Reducer，但它能有效减少 Mapper 和 Reducer 之间的数据传输量。

7.1.5　Reducer 类

Mapper 类有 4 个函数，在 Reducer 类中同样也有相应的 4 个函数：

protected void setup(Mapper.Context context) throws IOException,InterruptedException

protected void cleanup(Mapper.Context context)throws IOException,InterruptedException

protected void reduce(KEYIN key, VALUEIN value Reducer.Context context)throws IOException, InterruptedException

public void run(Reducer.Context context)throws IOException,InterruptedException

在用户的应用程序中调用到 Reducer 时，会直接调用 Reducer 里面的 run 函数，其代码如下：

```java
public void run(Context context) throws IOException, InterruptedException {
    setup(context);
    try {
        while (context.nextKey()) {
            reduce(context.getCurrentKey(), context.getValues(), context);
            // If a back up store is used, reset it
            Iterator<VALUEIN> iter = context.getValues().iterator();
            if(iter instanceof ReduceContext.ValueIterator) {
                ((ReduceContext.ValueIterator<VALUEIN>)iter).resetBackupStore();
            }
        }
    } finally {
        cleanup(context);
    }
}
```

可以看出，在 run 方法中调用了上面的 3 个方法：setup 方法、reduce 方法和 cleanup 方法。其中 setup 方法和 cleanup 方法默认是不做任何操作，且它们只被执行一次。但是，setup 方法一般会在 reduce 函数之前执行一些准备工作，如作业的一些配置信息等。cleanup 方法则是在 reduce 方法运行完之后最后执行的，该方法是完成一些结尾清理的工作，如资源释放等。如果需要做一些配置和清理的工作，需要在 Mapper/Reducer 的子类中进行重写来实现相应的功能。Reduce 方法会在对应的子类中重新实现，就是我们自定义的 reduce 方法，该方法在一个 while 循环里面，表明该方法是执行很多次的。run 方法就是每个 reduce task 调用的方法。

7.1.6 OutputFormat 接口

OutputFormat 主要用于描述输出数据的格式，它能够将用户提供的 key/value 对写入特定格式的文件中。Hadoop 自带了很多 OutputFormat 的实现，它们与 InputFormat 实现相对应，足够满足我们业务的需要。OutputFormat 是 MapReduce 输出的基类，所有 MapReduce 输出都实现了 OutputFormat 接口。我们可以把这些实现接口类分为几种类型来分别介绍。

1. 文本输出

默认的输出格式是 TextOutputFormat，它把每条记录写为文本行。它的键和值可以是任意类型，因为 TextOutputFormat 调用 toString()方法把它们转换为字符串。每个键/值对由"\t"进行分隔，当然也可以设定 mapreduce.output.textoutputformat.separator 属性改变默认的分隔符。与 TextOutputFormat 对应的输入格式是 KeyValueTextInputFormat，它通过可配置的分隔符将键/值对文本分隔。

可以使用 NullWritable 来省略输出的键或值（或两者都省略，相当于 NullOutputFormat 输出格式，后者什么也不输出）。这也会导致无分隔符输出，以使输出适合用 TextInputFormat 读取。

2. 二进制输出

（1）SequenceFileOutputFormat。

顾名思义，SequenceFileOutputFormat 将它的输出写为一个顺序文件。如果输出需要作为后续 MapReduce 任务的输入，这便是一种很好的输出格式，因为它的格式紧凑，很容易被压缩。

（2）SequenceFileAsBinaryOutputFormat。

SequenceFileAsBinaryOutputFormat 把键/值对作为二进制格式写到一个 SequenceFile 容器中。

（3）MapFileOutputFormat。

MapFileOutputFormat 把 MapFile 作为输出。MapFile 中的键必须顺序添加，所以必须确保 Reduce 输出的键已经排好序。

3. 多个输出

上面我们提到，默认情况下只有一个 Reduce，输出只有一个文件。有时可能需要对输出的文件名进行控制或让每个 Reduce 输出多个文件，我们有两种方式实现 Reduce 输出多个文件。

（1）Partitioner。

这种方法实现多文件输出，但很多情况下是无法实现的，因为存在以下两个缺点：

1）需要在作业运行之前知道分区数，如果分区数未知，则无法操作。

2）一般来说，让应用程序来严格限定分区数并不好，因为可能导致分区数少或分区不均。

（2）MultipleOutputs 类。

MultipleOutputs 类可以将数据写到多个文件中，这些文件的名称源于输出的键值或者任意字符串。允许每个 Reducer（或者只有 Map 作业的 Mapper）创建多个文件。采用 name-m-nnnnn 形式的文件名用于 Map 输出，name-r-nnnnn 形式的文件名用于 Reduce 输出，其中 name 是由程序设定的任意名字，nnnnn 是一个指明块号的整数（从 0 开始），块号保证从不同块（Mapper 或 Reducer）输出在相同名字情况下不会冲突。

4. 数据库输出

DBOutputFormat 适用于将作业输出数据（中等规模的数据）转存到 MySQL、Oracle 等数据库，使用 DBOutputFormat 以 MapReduce 的方式运行会并行地连接数据库。在这里需要合理地设置 Map、Reduce 的个数，以便将并行连接的数量控制在合理的范围之内。

7.1.7 GenericOptionsParser 类

GenericOptionsParser(Optionsopts,Configuration conf,String[] args)解析命令行参数。GenericOptionsParser 是为 Hadoop 框架解析命令行参数的工具类，它能够辨认标准的命令行参数，使程序能够轻松指定 NameNode、JobTracker，以及额外的配置资源或信息等。它支持的功能有：

- conf：指定配置文件。
- D：指定配置信息。
- fs：指定 NameNode。
- jt：指定 JobTracker。
- files：指定需要复制到 MapReduce 集群的文件，以逗号分隔。
- libjars：指定需要复制到 MapReduce 集群的 classpath 的 JAR 包，以逗号分隔。
- archives：指定需要复制到 MapReduce 集群的压缩文件，以逗号分隔，会自动解压缩。

7.1.8 DistributedCache 类

DistributedCache 可以用来分发简单的只读文件或者一些复杂的如 archive、jar 文件等。archive 文件会自动解压缩，而 jar 文件会被自动放置到任务的 classpath（lib）中。分发压缩 archive 时，可以指定解压名称，如 dict.zip#dict。这样就会解压到 dict 中，否则默认在 dict.zip 中。

程序通过 JobConf 中的 URI 来指定需要缓存的文件，它会假定指定的这个文件已经在 URI 指定的对应位置上了。程序在节点执行之前，DistributedCache 会复制必要的文件到这个 slave 节点。它的功效就是为每个程序只复制一次，而且复制到指定位置，能够自动解压缩。文件是有执行权限的。用户可以选择在任务的工作目录下建立指向 DistributedCache 的软链接。

DistributedCache.createSymlink(conf);
DistributedCache.addCacheFile(new Path("hdfs://host:port/absolute-path#link-name").toUri(), conf);

DistributedCache.createSymlink(Configuration)方法让 DistributedCache 在当前工作目录下创建到缓存文件的符号链接，在 task 的当前工作目录会有 link-name 的链接，相当于快捷方法，链接到 absolute-path 文件，在 setup 方法使用的情况则要简单许多。或者通过设置配置文件属性 mapred.create.symlink 为 yes，分布式缓存会截取 URI 的片段作为链接的名字，例如 URI 是 hdfs://host:port/lib.so.1#lib.so，则在 task 当前工作目录会有名为 lib.so 的链接，它会链接分布式缓存中的 lib.so.1。

DistributedCache 会跟踪修改缓存文件的 timestamp。下面是使用的例子，为应用程序设置缓存。

（1）将需要的文件复制到 FileSystem 中。

```
$ bin/hadoop fs -copyFromLocal lookup.dat /myapp/lookup.dat
$ bin/hadoop fs -copyFromLocal map.zip /myapp/map.zip
$ bin/hadoop fs -copyFromLocal mylib.jar /myapp/mylib.jar
$ bin/hadoop fs -copyFromLocal mytar.tar /myapp/mytar.tar
$ bin/hadoop fs -copyFromLocal mytgz.tgz /myapp/mytgz.tgz
$ bin/hadoop fs -copyFromLocal mytargz.tar.gz /myapp/mytargz.tar.gz
```

（2）设置 app 的 JobConf。

```
JobConf job = new JobConf();
DistributedCache.addCacheFile(new URI("/myapp/lookup.dat#lookup.dat"), job);
DistributedCache.addCacheArchive(new URI("/myapp/map.zip", job);
DistributedCache.addFileToClassPath(new Path("/myapp/mylib.jar"), job);
DistributedCache.addCacheArchive(new URI("/myapp/mytar.tar", job);
DistributedCache.addCacheArchive(new URI("/myapp/mytgz.tgz", job);
DistributedCache.addCacheArchive(new URI("/myapp/mytargz.tar.gz", job);
```

（3）在 Mapper 或 Reducer 中使用缓存文件。

```
public static class MapClass extends MapReduceBase
implements Mapper<K, V, K, V> {
    private Path[] localArchives;
    private Path[] localFiles;
    public void configure(JobConf job) {
        //得到刚刚缓存的文件
        localArchives = DistributedCache.getLocalCacheArchives(job);
        localFiles = DistributedCache.getLocalCacheFiles(job);
    }
    public void map(K key, V value,
    OutputCollector<K, V>; output, Reporter reporter)
    throws IOException {
        output.collect(k, v);
    }
}
```

7.2　MapReduce Java API 应用实例

7.2.1　统计单词出现频率

实现对指定目录或文件中的单词出现次数进行统计，默认输出结果是以单词字典排序。采用默认文本读入，每行读取一次，然后使用\t 对数据进行分割或者使用字符串类 StringTokenizer 对其分割（该类会按照空格、\t、\n 等进行切分）。在 Reduce 端相同的 key，即相同的单词会在一起进行求和处理，求出出现次数。数据集如图 7-2 所示。

图 7-2　统计单词频率数据集

实现代码如下：

```
import java.io.IOException;
import java.util.StringTokenizer;
import org.apache.hadoop.conf.Configuration;
import org.apache.hadoop.fs.Path;
import org.apache.hadoop.io.IntWritable;
import org.apache.hadoop.io.Text;
import org.apache.hadoop.mapreduce.Job;
import org.apache.hadoop.mapreduce.Mapper;
import org.apache.hadoop.mapreduce.Reducer;
import org.apache.hadoop.mapreduce.lib.input.FileInputFormat;
import org.apache.hadoop.mapreduce.lib.output.FileOutputFormat;

public class WordCount {

    //1.编写 map 函数，通过继承 Mapper 类实现里面的 map 函数
    //Mapper 类当中的第一个参数是 Object（常用），也可以写成 Long
    //第一个参数对应的值是行偏移量
    //2.第二个参数类型通常是 Text 类型，Text 是 Hadoop 实现的 String 类型的可写类型
    //第二个参数对应的值是每行字符串
    //第三个参数表示的是输出 key 的数据类型
    //第四个参数表示的是输出 value 的数据类型，IntWritable 是 Hadoop 实现的 int 类型的可写
    //数据类型
    public static class TokenizerMapper    extends Mapper<Object, Text, Text, IntWritable>{

        private final static IntWritable one = new IntWritable(1);
        private Text word = new Text();

        //key 是行偏移量
        //value 是每行字符串
        public void map(Object key, Text value, Context context) throws IOException, InterruptedException
        {
            StringTokenizer itr = new StringTokenizer(value.toString());
            while (itr.hasMoreTokens()) {
                word.set(itr.nextToken());          //itr.nextToken()是字符串类型，使用 set 函数完成字符串到 Text
                                                    //数据类型的转换
                context.write(word, one);           //通过 write 函数写入到本地文件
            }
        }
    }
```

```java
//第一个参数类型是输入值 key 的数据类型，map 中间输出 key 的数据类型
//第二个参数类型是输入值 value 的数据类型，map 中间输出 value 的数据类型
//第三个参数类型是输出值 key 的数据类型，它的数据类型要跟 job.setOutputKeyClass(Text.class)
//保持一致
//第四个参数类型是输出值 value 的数据类型，它的数据类型要跟 job.setOutputValueClass(IntWritable.class)
//保持一致
public static class IntSumReducer extends Reducer<Text,IntWritable,Text,IntWritable> {
    private IntWritable result = new IntWritable();
    //key 就是单词，values 是单词出现频率列表
    public void reduce(Text key, Iterable<IntWritable> values,
                       Context context
                       ) throws IOException, InterruptedException {
        int sum = 0;
        for (IntWritable val : values) {
            sum += val.get();           //get 就是取出 IntWritable 的值
        }
        result.set(sum);
        context.write(key, result);
    }
}

public static void main(String[] args) throws Exception {
    Configuration conf = new Configuration();
    //下面两种方式都是远程调试使用的（任选一种即可），如果说程序是在 Hadoop 集群运行的
    //则可以不用
    //conf.addResource(new Path("C:\\hadoop\\core-site.xml"));
    conf.set("fs.defaultFS", "hdfs://192.168.254.128:9000");

    //实例化一个作业，word count 是作业的名字
    Job job = Job.getInstance(conf, "word count");

    //指定通过哪个类找到对应的 JAR 包
    job.setJarByClass(WordCount.class);
    job.setMapperClass(TokenizerMapper.class);      //为 job 设置 Mapper 类
    //combiner 类也可以不需要
    //job.setCombinerClass(IntSumReducer.class);     //为 job 设置 Combiner 类
    job.setReducerClass(IntSumReducer.class);       //为 job 设置 Reducer 类
    job.setOutputKeyClass(Text.class);              //为 job 的输出数据设置 Key 类
    job.setOutputValueClass(IntWritable.class);     //为 job 输出设置 value 类

    //输入路径是存在的文件夹/文件
    FileInputFormat.addInputPath(job, new Path("/data/data1/1.txt"));   //为 job 设置输入路径
    //输出路径一定是不存在的文件夹
    FileOutputFormat.setOutputPath(job, new Path("/test2/wc01"));       //为 job 设置输出路径
    job.waitForCompletion(true);            //运行 job
    }
}
```

程序运行由于权限问题会报错，如图 7-3 所示。

```
18/01/16 16:34:38 WARN mapred.LocalJobRunner: job_local942097899_0001
org.apache.hadoop.security.AccessControlException: Permission denied: user=arp, access=WRITE, inode="/
    at org.apache.hadoop.hdfs.server.namenode.FSPermissionChecker.checkFsPermission(FSPermissionCh
    at org.apache.hadoop.hdfs.server.namenode.FSPermissionChecker.check(FSPermissionChecker.java:
    at org.apache.hadoop.hdfs.server.namenode.FSPermissionChecker.check(FSPermissionChecker.java:
    at org.apache.hadoop.hdfs.server.namenode.FSPermissionChecker.checkPermission(FSPermissionChe
    at org.apache.hadoop.hdfs.server.namenode.FSNamesystem.checkPermission(FSNamesystem.java:6547
    at org.apache.hadoop.hdfs.server.namenode.FSNamesystem.checkPermission(FSNamesystem.java:6529
    at org.apache.hadoop.hdfs.server.namenode.FSNamesystem.checkAncestorAccess(FSNamesystem.java:6
    at org.apache.hadoop.hdfs.server.namenode.FSNamesystem.mkdirsInternal(FSNamesystem.java:4290)
    at org.apache.hadoop.hdfs.server.namenode.FSNamesystem.mkdirsInt(FSNamesystem.java:4260)
    at org.apache.hadoop.hdfs.server.namenode.FSNamesystem.mkdirs(FSNamesystem.java:4233)
    at org.apache.hadoop.hdfs.server.namenode.NameNodeRpcServer.mkdirs(NameNodeRpcServer.java:853)
    at org.apache.hadoop.hdfs.protocolPB.ClientNamenodeProtocolServerSideTranslatorPB.mkdirs(Clie
    at org.apache.hadoop.hdfs.protocol.proto.ClientNamenodeProtocolProtos$ClientNamenodeProtocol$2
    at org.apache.hadoop.ipc.ProtobufRpcEngine$Server$ProtoBufRpcInvoker.call(ProtobufRpcEngine.ja
    at org.apache.hadoop.ipc.RPC$Server.call(RPC.java:975)
    at org.apache.hadoop.ipc.Server$Handler$1.run(Server.java:2040)
    at org.apache.hadoop.ipc.Server$Handler$1.run(Server.java:2036)
    at java.security.AccessController.doPrivileged(Native Method)
```

图 7-3　权限错误提示

对文件/test2 修改权限为：hdfs dfs -chmod 777 /test2。

运行成功会有如图 7-4 所示的文件存在，_SUCCESS 表示运行成功，part-r-00000 存放的是运行结果，由于默认设置为一个分区，因此文件后置是 00000，如果设置多个，则后缀依次是 00001、00002 等。

图 7-4　WordCount 运行结果

可以看到 part-r-00000 的文件内容如图 7-5 所示。

图 7-5　part-r-00000 输出内容

为了看到 MapReduce 执行情况，可以将 Hadoop 中的 etc/hadoop/log4j.properties 文件拷贝到 src 目录下，此时控制台会打印相应输出信息，如图 7-6 所示。

```
18/01/20 10:42:51 INFO mapred.LocalJobRunner: 1 / 1 copied.
18/01/20 10:42:51 INFO Configuration.deprecation: mapred.skip.on is deprecated. Instead, use mapreduce.job.s
18/01/20 10:42:52 INFO mapreduce.Job: Job job_local965908676_0001 running in uber mode : false
18/01/20 10:42:52 INFO mapreduce.Job:  map 100% reduce 0%
18/01/20 10:42:52 INFO mapred.Task: Task:attempt_local965908676_0001_r_000000_0 is done. And is in the proce
18/01/20 10:42:52 INFO mapred.LocalJobRunner: 1 / 1 copied.
18/01/20 10:42:52 INFO mapred.Task: Task attempt_local965908676_0001_r_000000_0 is allowed to commit now
18/01/20 10:42:52 INFO output.FileOutputCommitter: Saved output of task 'attempt_local965908676_0001_r_00000
18/01/20 10:42:52 INFO mapred.LocalJobRunner: reduce > reduce
18/01/20 10:42:52 INFO mapred.Task: Task 'attempt_local965908676_0001_r_000000_0' done.
18/01/20 10:42:52 INFO mapred.LocalJobRunner: Finishing task: attempt_local965908676_0001_r_000000_0
18/01/20 10:42:52 INFO mapred.LocalJobRunner: reduce task executor complete.
18/01/20 10:42:53 INFO mapreduce.Job:  map 100% reduce 100%
18/01/20 10:42:53 INFO mapreduce.Job: Job job_local965908676_0001 completed successfully
18/01/20 10:42:53 INFO mapreduce.Job: Counters: 35
        File System Counters
                FILE: Number of bytes read=648
                FILE: Number of bytes written=504261
                FILE: Number of read operations=0
                FILE: Number of large read operations=0
                FILE: Number of write operations=0
                HDFS: Number of bytes read=138
                HDFS: Number of bytes written=56
                HDFS: Number of read operations=13
                HDFS: Number of large read operations=0
                HDFS: Number of write operations=4
        Map-Reduce Framework
                Map input records=5
                Map output records=11
                Map output bytes=113
                Map output materialized bytes=141
                Input split bytes=109
                Combine input records=0
                Combine output records=0
                Reduce input groups=7
```

图 7-6 MapReduce 日志输出

导出 JAR 包部署到集群运行和 HDFS 方法一样，这里不再过多描述。

7.2.2 统计出现的单词

与统计出现单词频率相比，只是将最终结果的出现次数去掉，因此只需将 WordCount 代码 reduce 函数中输出 value 的值设置为 NullWritable 即可，同时将 Job 的 OutputValue 类型设置为 NullWritable，数据集如图 7-7 所示。

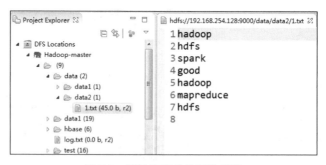

图 7-7 统计出现的单词数据集

实现代码如下：

```
import java.io.IOException;
import java.util.StringTokenizer;
import org.apache.hadoop.conf.Configuration;
import org.apache.hadoop.fs.Path;
import org.apache.hadoop.io.IntWritable;
```

```java
import org.apache.hadoop.io.NullWritable;
import org.apache.hadoop.io.Text;
import org.apache.hadoop.mapreduce.Job;
import org.apache.hadoop.mapreduce.Mapper;
import org.apache.hadoop.mapreduce.Reducer;
import org.apache.hadoop.mapreduce.lib.input.FileInputFormat;
import org.apache.hadoop.mapreduce.lib.output.FileOutputFormat;

public class WordDelete {

    //去除文件中单词的重复内容
    //跟统计单词内容相比，只是最后的 reduce 函数写入文件内容的 value 值变化

    public static class TokenizerMapper    extends Mapper<Object, Text, Text, IntWritable>{


        private final static IntWritable one = new IntWritable(1);
        private Text word = new Text();

        //key 是行偏移量
        //value 是每行字符串
        public void map(Object key, Text value, Context context) throws IOException, InterruptedException
        {
            StringTokenizer itr = new StringTokenizer(value.toString());
            while (itr.hasMoreTokens()) {
                word.set(itr.nextToken());       //itr.nextToken()是字符串类型,使用 set 函数完成字符串到 Text
                                                 //数据类型的转换
                context.write(word, one);        //通过 write 函数写入到本地文件
            }
        }
    }

    public static class IntSumReducer extends Reducer<Text,IntWritable,Text,NullWritable> {
        private IntWritable result = new IntWritable();
        //key 就是单词，values 是单词出现频率列表
        public void reduce(Text key, Iterable<IntWritable> values,
                           Context context
                           ) throws IOException, InterruptedException {
//          int sum = 0;
//          for (IntWritable val : values) {
//              sum += val.get();        //get 就是取出 val 的值
//          }
//          result.set(sum);
            NullWritable a = null;
            context.write(key, a);
        }
    }
```

```java
    public static void main(String[] args) throws Exception {
        Configuration conf = new Configuration();
        //下面两种方式都是远程调试使用的（任选一种即可），如果说程序是在hadoop集群运行的，则
        //可以不用
//      conf.addResource(new Path("C:\\hadoop\\core-site.xml"));
        conf.set("fs.defaultFS", "hdfs://192.168.254.128:9000");

        //实例化一个作业，word delete 是作业的名字
        Job job = Job.getInstance(conf, "word delete");

        //指定通过哪个类找到对应的 JAR 包
        job.setJarByClass(WordDelete.class);
        job.setMapperClass(TokenizerMapper.class);        //为 job 设置 Mapper 类
        //combiner 类也可以不需要
        //job.setCombinerClass(IntSumReducer.class);      //为 job 设置 Combiner 类
        job.setReducerClass(IntSumReducer.class);         //为 job 设置 Reducer 类

        //设置 map 的输出 value 的数据类型
        job.setMapOutputValueClass(IntWritable.class);
        job.setOutputKeyClass(Text.class);                //为 job 的输出数据设置 key 类
        job.setOutputValueClass(NullWritable.class);      //为 job 的输出设置 value 类

        //输入路径是存在的文件夹/文件
        FileInputFormat.addInputPath(job, new Path(args[0]));    //为 job 设置输入路径
        //输出路径一定是不存在的文件夹
        FileOutputFormat.setOutputPath(job, new Path(args[1]));  //为 job 设置输出路径
        job.waitForCompletion(true);                             //运行 job
    }
}
```

最终的输出结果如图 7-8 所示。

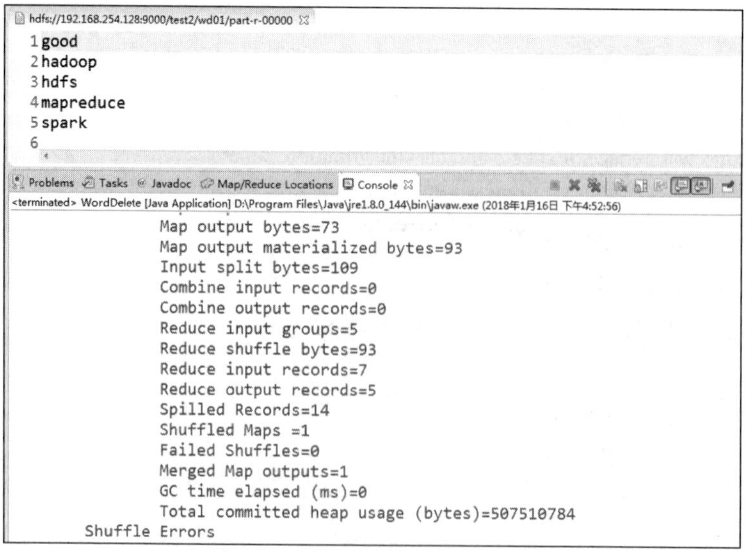

图 7-8　WordDelete 输出结果

7.2.3 统计平均成绩

统计学生各科平均成绩，每科成绩为一个文件。在 Map 阶段和 WordCount 一样，只是在 Reduce 阶段求出总和之后，再除以科目数，并将输出 value 的数据类型设置为 FloatWritable 即可。数据集如图 7-9 至图 7-11 所示，第一列是姓名，第二列是成绩。

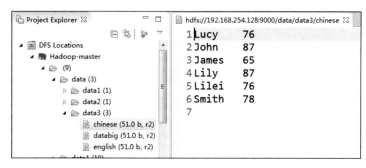

图 7-9　chinese 成绩数据集

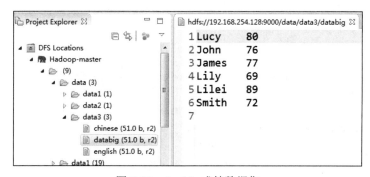

图 7-10　databig 成绩数据集

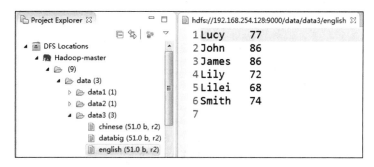

图 7-11　english 成绩数据集

实现代码如下：

```
import java.io.IOException;
import java.util.StringTokenizer;
import org.apache.hadoop.conf.Configuration;
import org.apache.hadoop.fs.Path;
import org.apache.hadoop.io.FloatWritable;
import org.apache.hadoop.io.IntWritable;
```

```java
import org.apache.hadoop.io.Text;
import org.apache.hadoop.mapreduce.Job;
import org.apache.hadoop.mapreduce.Mapper;
import org.apache.hadoop.mapreduce.Reducer;
import org.apache.hadoop.mapreduce.lib.input.FileInputFormat;
import org.apache.hadoop.mapreduce.lib.output.FileOutputFormat;

public class ScoreAverage {
  public static class TokenizerMapper    extends Mapper<Object, Text, Text, IntWritable>{

    private final static IntWritable one = new IntWritable(1);
    private Text word = new Text();

    public void map(Object key, Text value, Context context) throws IOException, InterruptedException
    {

        String[] str = value.toString().split("\t");
        context.write(new Text(str[0]), new IntWritable(Integer.parseInt(str[1])));
    }
  }

  public static class IntSumReducer extends Reducer<Text,IntWritable,Text,FloatWritable> {
    private FloatWritable result = new FloatWritable();
    //与 WordCount 相比就是在 value 的设置时除以文件个数，如下面的 3
    public void reduce(Text key, Iterable<IntWritable> values,
                       Context context
                       ) throws IOException, InterruptedException {
      int sum = 0;
      for (IntWritable val : values) {
        sum += val.get();        //get 就是取出 IntWritable 的值
      }
       //3 表示科目数，本实例中每个科目对应一个文件
      result.set((float)sum / 3);
      context.write(key, result);
    }
  }

  public static void main(String[] args) throws Exception {
    Configuration conf = new Configuration();
    //下面两种方式都是远程调试使用的（任选一种即可），如果说程序是在 hadoop 集群运行的
    //则可以不用
    //    conf.addResource(new Path("C:\\hadoop\\core-site.xml"));
 conf.set("fs.defaultFS", "hdfs://192.168.254.128:9000");

    //实例化一个作业，ScoreAverage 是作业的名字
    Job job = Job.getInstance(conf, "ScoreAverage");
```

//指定通过哪个类找到对应的 JAR 包
job.setJarByClass(ScoreAverage.class);
job.setMapperClass(TokenizerMapper.class);   //为 job 设置 Mapper 类
job.setReducerClass(IntSumReducer.class);   //为 job 设置 Reducer 类
job.setMapOutputValueClass(IntWritable.class);
job.setOutputKeyClass(Text.class);   //为 job 的输出数据设置 Key 类
job.setOutputValueClass(FloatWritable.class);   //为 job 输出设置 value 类

//输入路径是存在的文件夹/文件
FileInputFormat.addInputPath(job, new Path("/data/data3"));   //为 job 设置输入路径
//输出路径一定是不存在的文件夹
FileOutputFormat.setOutputPath(job, new Path("/test2/sc01")); //为 job 设置输出路径
job.waitForCompletion(true);   //运行 job
   }
 }
```

最终的输出结果如图 7-12 所示。

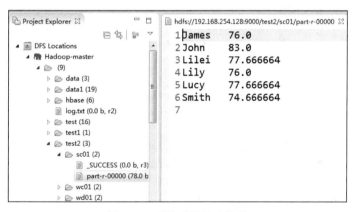

图 7-12　平均成绩输出结果

### 7.2.4　排序

输入数据格式为每行有一数值，通过 MapReduce 实现数据的排序功能。利用 Map 阶段的 Sort 功能将要排序的数值作为 map 函数的 key 输出，并在 reduce 函数设置一个计数器。数据集如图 7-13 所示。

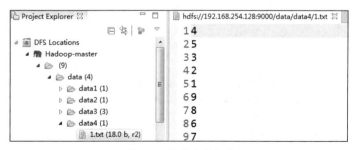

图 7-13　待排序数据集
```

实现代码如下：

```java
import java.io.IOException;
import org.apache.hadoop.conf.Configuration;
import org.apache.hadoop.fs.Path;
import org.apache.hadoop.io.IntWritable;
import org.apache.hadoop.io.Text;
import org.apache.hadoop.mapreduce.Job;
import org.apache.hadoop.mapreduce.Mapper;
import org.apache.hadoop.mapreduce.Reducer;
import org.apache.hadoop.mapreduce.lib.input.FileInputFormat;
import org.apache.hadoop.mapreduce.lib.output.FileOutputFormat;
import org.apache.hadoop.util.GenericOptionsParser;

public class Sort {
    //map 将输入中的 value 化成 IntWritable 类型，作为输出的 key
    public static class Map extends Mapper<Object,Text,IntWritable,IntWritable>{
        private static IntWritable data=new IntWritable();
        //实现 map 函数
        public void map(Object key,Text value,Context context)
                throws IOException,InterruptedException{
            String line=value.toString();
            data.set(Integer.parseInt(line));
            context.write(data, new IntWritable(1));
        }
    }
    //reduce 将输入中的 key 复制到输出数据的 key 上
    //然后根据输入的 value-list 中元素的个数决定 key 的输出次数
    //利用了 shuffle 对 key 的排序原理，故 map 中的 key 为数据
    public static class Reduce extends
            Reducer<IntWritable,IntWritable,IntWritable,IntWritable>{
        private static IntWritable linenum = new IntWritable(1);
        private static int i = 1;
        //实现 reduce 函数
        public void reduce(IntWritable key,Iterable<IntWritable> values,Context context)
                throws IOException,InterruptedException{

            context.write(new IntWritable(i), key);
            ++i;
        }
    }

    public static void main(String[] args) throws Exception{
        Configuration conf = new Configuration();
        conf.set("fs.defaultFS", "hdfs://192.168.254.128:9000");
        Job job = Job.getInstance(conf, "sort");
        job.setJarByClass(Sort.class);
```

//设置 Map 和 Reduce 处理类
job.setMapperClass(Map.class);
job.setReducerClass(Reduce.class);

//设置输出类型
job.setOutputKeyClass(IntWritable.class);
job.setOutputValueClass(IntWritable.class);

//设置输入和输出目录
FileInputFormat.addInputPath(job, new Path("/data/data4"));
FileOutputFormat.setOutputPath(job, new Path("/test2/sort01"));
System.exit(job.waitForCompletion(true) ? 0 : 1);
    }
}

最终的输出结果如图 7-14 所示。

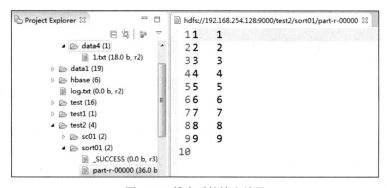

图 7-14　排序后的输出结果

7.2.5　求年最高温度

输入数据为每条一行，每条数据包含年、月、日、温度。在 Map 阶段将年作为输出的 key，温度作为输出的 value；在 Reduce 阶段求出相同 key 的 value 最大值。数据集如图 7-15 所示，前四位数据为年，接下来两位是月份，再后面两位是日期，最后两位是气温。

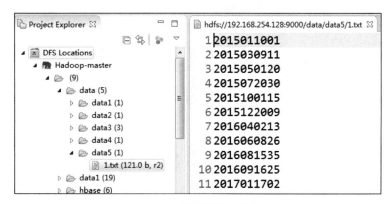

图 7-15　温度统计数据集

实现代码如下：

```java
import java.io.IOException;
import java.util.StringTokenizer;
import org.apache.hadoop.conf.Configuration;
import org.apache.hadoop.fs.Path;
import org.apache.hadoop.io.IntWritable;
import org.apache.hadoop.io.Text;
import org.apache.hadoop.mapreduce.Job;
import org.apache.hadoop.mapreduce.Mapper;
import org.apache.hadoop.mapreduce.Reducer;
import org.apache.hadoop.mapreduce.lib.input.FileInputFormat;
import org.apache.hadoop.mapreduce.lib.output.FileOutputFormat;

public class Temperature {

    public static class TokenizerMapper    extends Mapper<Object, Text,
            Text, IntWritable>{
        private final static IntWritable one = new IntWritable(1);
        private Text word = new Text();

        //key 是行偏移量
        //value 是每行字符串
        public void map(Object key, Text value, Context context) throws IOException, InterruptedException
        {
            String year = value.toString().substring(0, 4);//[0,4)
            int temp = Integer.parseInt(value.toString().substring(8));
            context.write(new Text(year), new IntWritable(temp));
        }
    }

    public static class IntSumReducer extends Reducer<Text,IntWritable,Text,IntWritable> {
        private IntWritable result = new IntWritable();
        public void reduce(Text key, Iterable<IntWritable> values,
                           Context context
                           ) throws IOException, InterruptedException {

            int max = Integer.MIN_VALUE;
            for(IntWritable val : values){
             if(max < val.get())
                    max = val.get();
            }

            context.write(key, new IntWritable(max));
        }
    }
```

```java
public static void main(String[] args) throws Exception {
    Configuration conf = new Configuration();
    //下面两种方式都是远程调试使用的（任选一种即可），如果说程序是在Hadoop集群运行的
    //则可以不用
    //    conf.addResource(new Path("C:\\hadoop\\core-site.xml"));
    conf.set("fs.defaultFS", "hdfs://192.168.254.128:9000");

    //实例化一个作业，word count 是作业的名字
    Job job = Job.getInstance(conf, "word count");

    //指定通过哪个类找到对应的jar包
    job.setJarByClass(Temperature.class);
    job.setMapperClass(TokenizerMapper.class);       //为job 设置 Mapper 类
    job.setReducerClass(IntSumReducer.class);        //为job 设置 Reducer 类
    job.setOutputKeyClass(Text.class);               //为job 的输出数据设置 Key 类
    job.setOutputValueClass(IntWritable.class);      //为job 输出设置 value 类

    //输入路径是存在的文件夹/文件
    FileInputFormat.addInputPath(job, new Path("/data/data5"));   //为job 设置输入路径
    //输出路径一定是不存在的文件夹
    FileOutputFormat.setOutputPath(job, new Path("/test2/tt01")); //为job 设置输出路径
    job.waitForCompletion(true);            //运行job
}
}
```

最终的输出结果如图 7-16 所示。

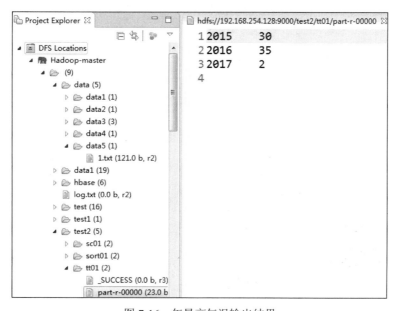

图 7-16　年最高气温输出结果

扩展：如果还要显示出最高温度的年月日，则需要将 Map 阶段的输出 value 设置为整条记录，在 Reduce 阶段首先求出温度再进行比较。

7.2.6 关系运算——投影运算

实现类似关系代数集合的投影运算,每条数据占一行,字段之间由"\t"分隔。只需要将要选择的字段放入 Map 阶段的输出 key,并在 Reduce 中将输出 value 设置为 NullWritable。数据集如图 7-17 和图 7-18 所示,第一、二、三列分别表示姓名、性别、年龄。

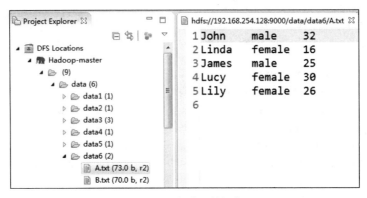

图 7-17 关系 A 数据集

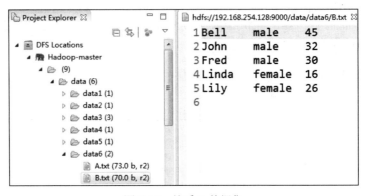

图 7-18 关系 B 数据集

实现代码如下:

```
import java.io.IOException;
import java.util.StringTokenizer;
import org.apache.hadoop.conf.Configuration;
import org.apache.hadoop.fs.Path;
import org.apache.hadoop.io.FloatWritable;
import org.apache.hadoop.io.IntWritable;
import org.apache.hadoop.io.NullWritable;
import org.apache.hadoop.io.Text;
import org.apache.hadoop.mapreduce.Job;
import org.apache.hadoop.mapreduce.Mapper;
import org.apache.hadoop.mapreduce.Reducer;
import org.apache.hadoop.mapreduce.lib.input.FileInputFormat;
import org.apache.hadoop.mapreduce.lib.output.FileOutputFormat;
```

```java
public class Select {
    public static class TokenizerMapper    extends Mapper<Object, Text, Text, IntWritable>{

        public void map(Object key, Text value, Context context) throws IOException, InterruptedException
        {

            String[] str = value.toString().split("\t");
            context.write(new Text(str[1]), new IntWritable(1));
        }
    }

    public static class IntSumReducer extends Reducer<Text,IntWritable,Text,NullWritable> {
        private FloatWritable result = new FloatWritable();
        public void reduce(Text key, Iterable<IntWritable> values,
                           Context context
                           ) throws IOException, InterruptedException {
            NullWritable a = null;
            context.write(key, a);
        }
    }

    public static void main(String[] args) throws Exception {
        Configuration conf = new Configuration();
        //下面两种方式都是远程调试使用的（任选一种即可），如果说程序是在 Hadoop 集群运行的
        //则可以不用
//        conf.addResource(new Path("C:\\hadoop\\core-site.xml"));
        conf.set("fs.defaultFS", "hdfs://192.168.254.128:9000");

        //实例化一个作业，Select 是作业的名字
        Job job = Job.getInstance(conf, "Select");

        //指定通过哪个类找到对应的 JAR 包
        job.setJarByClass(Select.class);
        job.setMapperClass(TokenizerMapper.class);         //为 job 设置 Mapper 类
        job.setReducerClass(IntSumReducer.class);          //为 job 设置 Reducer 类

        job.setMapOutputValueClass(IntWritable.class);
        job.setOutputKeyClass(Text.class);                 //为 job 的输出数据设置 Key 类
        job.setOutputValueClass(NullWritable.class);       //为 job 输出设置 value 类

        //输入路径是存在的文件夹/文件
        FileInputFormat.addInputPath(job, new Path("/data/data6"));    //为 job 设置输入路径
        //输出路径一定是不存在的文件夹
        FileOutputFormat.setOutputPath(job, new Path("/test2/rel01")); //为 job 设置输出路径
        job.waitForCompletion(true);           //运行 job
    }
}
```

最终的输出结果如图 7-19 所示。

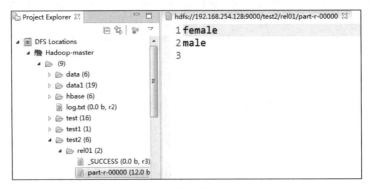

图 7-19　投影运算输出结果

7.2.7　关系运算——并运算

实现类似关系代数集合的并运算，每条数据占一行，字段之间由"\t"分隔。在 Map 阶段将整条记录作为输出的 key，在 Reduce 阶段直接输出即可，数据集同投影运算。

实现代码如下：

```
import java.io.IOException;
import java.util.StringTokenizer;
import org.apache.hadoop.conf.Configuration;
import org.apache.hadoop.fs.Path;
import org.apache.hadoop.io.FloatWritable;
import org.apache.hadoop.io.IntWritable;
import org.apache.hadoop.io.NullWritable;
import org.apache.hadoop.io.Text;
import org.apache.hadoop.mapreduce.Job;
import org.apache.hadoop.mapreduce.Mapper;
import org.apache.hadoop.mapreduce.Reducer;
import org.apache.hadoop.mapreduce.lib.input.FileInputFormat;
import org.apache.hadoop.mapreduce.lib.output.FileOutputFormat;

public class Union {
    public static class TokenizerMapper    extends Mapper<Object, Text, Text, IntWritable>{

        public void map(Object key, Text value, Context context) throws IOException, InterruptedException
        {

            context.write(value, new IntWritable(1));
        }
    }

    public static class IntSumReducer extends Reducer<Text,IntWritable,Text,NullWritable> {
      private FloatWritable result = new FloatWritable();
      //与 WordCount 相比就是在 value 的设置时除以文件个数，如下面的 3
```

```java
        public void reduce(Text key, Iterable<IntWritable> values,
                           Context context
                          ) throws IOException, InterruptedException {
            NullWritable a = null;
            context.write(key, a);
        }
    }

    public static void main(String[] args) throws Exception {
        Configuration conf = new Configuration();
        //下面两种方式都是远程调试使用的（任选一种即可），如果说程序是在 Hadoop 集群运行的
        //则可以不用
//      conf.addResource(new Path("C:\\hadoop\\core-site.xml"));
        conf.set("fs.defaultFS", "hdfs://192.168.254.128:9000");

        //实例化一个作业，Union 是作业的名字
        Job job = Job.getInstance(conf, "Union");

        //指定通过哪个类找到对应的 JAR 包
        job.setJarByClass(Union.class);
        job.setMapperClass(TokenizerMapper.class);           //为 job 设置 Mapper 类
        job.setReducerClass(IntSumReducer.class);            //为 job 设置 Reducer 类

        job.setMapOutputValueClass(IntWritable.class);
        job.setOutputKeyClass(Text.class);                   //为 job 的输出数据设置 Key 类
        job.setOutputValueClass(NullWritable.class);         //为 job 输出设置 value 类

        //输入路径是存在的文件夹/文件
        FileInputFormat.addInputPath(job, new Path("/data/data6"));     //为 job 设置输入路径
        //输出路径一定是不存在的文件夹
        FileOutputFormat.setOutputPath(job, new Path("/test2/rel02"));  //为 job 设置输出路径
        job.waitForCompletion(true);             //运行 job
    }
}
```
最终的输出结果如图 7-20 所示。

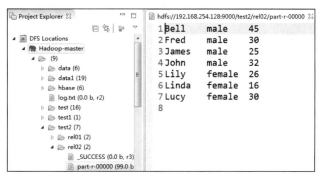

图 7-20　并运算输出结果

7.2.8 关系运算——交运算

实现类似关系代数集合的交运算,每条数据占一行,字段之间由 "\t" 分隔。在 Map 阶段将整条记录作为输出的 key,在 Reduce 阶段对相同 key 的 value 进行计算,凡是大于等于 2 的为符合条件的,将其保存,数据集同投影运算。

实现代码如下:

```java
import java.io.IOException;
import java.util.StringTokenizer;
import org.apache.hadoop.conf.Configuration;
import org.apache.hadoop.fs.Path;
import org.apache.hadoop.io.FloatWritable;
import org.apache.hadoop.io.IntWritable;
import org.apache.hadoop.io.NullWritable;
import org.apache.hadoop.io.Text;
import org.apache.hadoop.mapreduce.Job;
import org.apache.hadoop.mapreduce.Mapper;
import org.apache.hadoop.mapreduce.Reducer;
import org.apache.hadoop.mapreduce.lib.input.FileInputFormat;
import org.apache.hadoop.mapreduce.lib.output.FileOutputFormat;

public class Intersect {
    public static class TokenizerMapper    extends Mapper<Object, Text, Text, IntWritable>{

        public void map(Object key, Text value, Context context) throws IOException, InterruptedException
        {

            context.write(value, new IntWritable(1));
        }
    }

    public static class IntSumReducer extends Reducer<Text,IntWritable,Text,NullWritable> {
        private FloatWritable result = new FloatWritable();
        public void reduce(Text key, Iterable<IntWritable> values,
                            Context context
                            ) throws IOException, InterruptedException {
            int sum = 0;
            for(IntWritable val : values){
             sum ++;
            }
            NullWritable a = null;
            if(sum >= 2){
             context.write(key, a);
            }
        }
    }
```

```
public static void main(String[] args) throws Exception {
    Configuration conf = new Configuration();
    //下面两种方式都是远程调试使用的（任选一种即可），如果说程序是在 hadoop 集群运行的
    //则可以不用
//  conf.addResource(new Path("C:\\hadoop\\core-site.xml"));
    conf.set("fs.defaultFS", "hdfs://192.168.254.128:9000");

    //实例化一个作业，Intersection 是作业的名字
    Job job = Job.getInstance(conf, "Intersection");

    //指定通过哪个类找到对应的 JAR 包
    job.setJarByClass(Intersect.class);
    job.setMapperClass(TokenizerMapper.class);       //为 job 设置 Mapper 类
    job.setReducerClass(IntSumReducer.class);        //为 job 设置 Reducer 类

    job.setMapOutputValueClass(IntWritable.class);
    job.setOutputKeyClass(Text.class);               //为 job 的输出数据设置 Key 类
    job.setOutputValueClass(NullWritable.class);     //为 job 输出设置 value 类

    //输入路径是存在的文件夹/文件
    FileInputFormat.addInputPath(job, new Path("/data/data6"));   //为 job 设置输入路径
    //输出路径一定是不存在的文件夹
    FileOutputFormat.setOutputPath(job, new Path("/test2/rel03")); //为 job 设置输出路径
    job.waitForCompletion(true);            //运行 job
  }
}
```

最终的输出结果如图 7-21 所示。

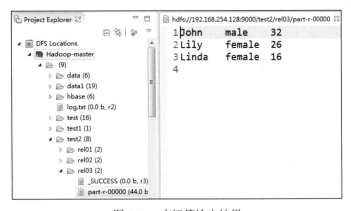

图 7-21　交运算输出结果

7.2.9　关系运算——差运算

实现类似关系代数集合的差运算，每条数据占一行，字段之间由 "\t" 分隔。在 Map 阶段将整条记录作为输出的 key，将文件名作为输出的 value；在 Reduce 阶段根据文件名判断，如果为被减的集合，则将结果输出，数据集同投影运算。

实现代码如下:

```java
import java.io.IOException;
import java.util.StringTokenizer;
import org.apache.hadoop.conf.Configuration;
import org.apache.hadoop.fs.Path;
import org.apache.hadoop.io.FloatWritable;
import org.apache.hadoop.io.IntWritable;
import org.apache.hadoop.io.NullWritable;
import org.apache.hadoop.io.Text;
import org.apache.hadoop.mapreduce.Job;
import org.apache.hadoop.mapreduce.Mapper;
import org.apache.hadoop.mapreduce.Reducer;
import org.apache.hadoop.mapreduce.lib.input.FileInputFormat;
import org.apache.hadoop.mapreduce.lib.input.FileSplit;
import org.apache.hadoop.mapreduce.lib.output.FileOutputFormat;

public class Minus {
    public static class TokenizerMapper    extends Mapper<Object, Text, Text, Text>{

        public void map(Object key, Text value, Context context) throws IOException, InterruptedException
        {
            FileSplit file = (FileSplit)context.getInputSplit();
            String fileName = file.getPath().getName();

            context.write(value, new Text(fileName));
        }
    }

    public static class IntSumReducer extends Reducer<Text,Text,Text,NullWritable> {
        private FloatWritable result = new FloatWritable();
        public void reduce(Text key, Iterable<Text> values,
                           Context context
                           ) throws IOException, InterruptedException {
            int flag = 0;
            for(Text val : values){
             if("B.txt".equals(val.toString())){
                    flag = 1;
                    break;
              }
            }

            if(flag == 0){
             NullWritable a = null;
                context.write(key, a);
            }
```

        }
    }

    public static void main(String[] args) throws Exception {
        Configuration conf = new Configuration();
        //下面两种方式都是远程调试使用的（任选一种即可），如果说程序是在 Hadoop 集群运行的
        //则可以不用
//   conf.addResource(new Path("C:\\hadoop\\core-site.xml"));
        conf.set("fs.defaultFS", "hdfs://192.168.254.128:9000");

        //实例化一个作业，minus 是作业的名字
        Job job = Job.getInstance(conf, "minus");

        //指定通过哪个类找到对应的 JAR 包
        job.setJarByClass(Minus.class);
        job.setMapperClass(TokenizerMapper.class);   //为 job 设置 Mapper 类
        job.setReducerClass(IntSumReducer.class);   //为 job 设置 Reducer 类

        job.setMapOutputValueClass(Text.class);
        job.setOutputKeyClass(Text.class);   //为 job 的输出数据设置 Key 类
        job.setOutputValueClass(NullWritable.class);   //为 job 输出设置 value 类

        //输入路径是存在的文件夹/文件
        FileInputFormat.addInputPath(job, new Path("/data/data6"));   //为 job 设置输入路径
        //输出路径一定是不存在的文件夹
        FileOutputFormat.setOutputPath(job, new Path("/test2/rel04")); //为 job 设置输出路径
        job.waitForCompletion(true);   //运行 job
    }
}

最终的输出结果如图 7-22 所示。

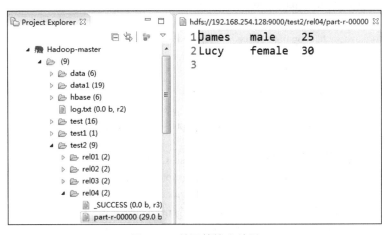

图 7-22　差运算输出结果

7.2.10 关系运算——连接运算

利用单表关联在父子关系中求解爷孙关系，有如图 7-23 所示的数据，设定左边是右边的儿子，右边是左边的父辈。在 Map 阶段，将父子关系与相反的子父关系同时在各个 value 前补上前缀-与+标识此 key-value 中的 value 是正序还是逆序产生的，之后进入 context。MapReduce 会自动将同一个 key 的不同的 value 值组合在一起推到 Reduce 阶段。在 value 数组中，根据前缀，我们可以轻松得知，哪个是爷爷，哪个是孙子。因此，对各个 values 数组中各个项的前缀进行输出。可以看出，整个过程 Key 一直被作为连接的桥梁使用，形成一个单表关联的运算。

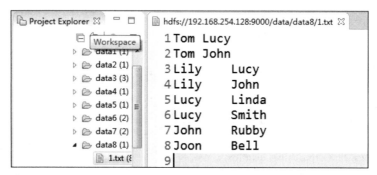

图 7-23　连接运算数据集

实现代码如下：

```
import java.io.IOException;
import java.util.ArrayList;
import org.apache.hadoop.conf.Configuration;
import org.apache.hadoop.fs.FileSystem;
import org.apache.hadoop.fs.Path;
import org.apache.hadoop.io.Text;
import org.apache.hadoop.mapreduce.Job;
import org.apache.hadoop.mapreduce.Mapper;
import org.apache.hadoop.mapreduce.Reducer;
import org.apache.hadoop.mapreduce.lib.input.FileInputFormat;
import org.apache.hadoop.mapreduce.lib.output.FileOutputFormat;
import org.apache.hadoop.util.GenericOptionsParser;

public class LeftOuterRightDemo {

    public static class MyMapper extends Mapper<Object, Text, Text, Text> {

        public void map(Object key, Text value, Context context)
                throws IOException, InterruptedException {
            String child = value.toString().split("\t")[0];
            String parent = value.toString().split("\t")[1];
            //产生正序与逆序的 key-value，同时压入 context
            context.write(new Text(child), new Text("-" + parent));
```

```java
                context.write(new Text(parent), new Text("+" + child));
            }
        }

        public static class MyReducer extends Reducer<Text, Text, Text, Text> {

            public void reduce(Text key, Iterable<Text> values, Context context)
                    throws IOException, InterruptedException {
                ArrayList<Text> grandparent = new ArrayList<Text>();
                ArrayList<Text> grandchild = new ArrayList<Text>();
                for (Text t : values) {//对各个 values 中的值进行处理
                    String s = t.toString();
                    if (s.startsWith("-")) {
                        grandparent.add(new Text(s.substring(1)));
                    } else {
                        grandchild.add(new Text(s.substring(1)));
                    }
                }
                //再将 grandparent 与 grandchild 中的东西一一对应输出
                for (int i = 0; i < grandchild.size(); i++) {
                    for (int j = 0; j < grandparent.size(); j++) {
                        context.write(grandchild.get(i), grandparent.get(j));
                    }
                }
            }
        }

        public static void main(String[] args) throws Exception {
            Configuration conf = new Configuration();
            conf.set("fs.defaultFS", "hdfs://192.168.254.128:9000");

            //如果是在 Hadoop 集群中运行，可以将下面的注释去掉
            //如果 Windows 远程调试，则必须将下面的代码注释掉
//          String[] otherArgs = new GenericOptionsParser(conf, args)
//                  .getRemainingArgs();
//          if (otherArgs.length != 2) {
//              System.err.println("Usage: wordcount <in> <out>");
//              System.exit(2);
//          }
            Job job = Job.getInstance(conf, "LeftOuterRightDemo");
            job.setMapperClass(MyMapper.class);
            job.setReducerClass(MyReducer.class);
            job.setOutputKeyClass(Text.class);
            job.setOutputValueClass(Text.class);

            // 判断 output 文件夹是否存在，如果存在则删除
```

```
//        Path path = new Path(otherArgs[1]);      //取第 1 个表示输出目录参数（第 0 个参数是输入目录）
//        FileSystem fileSystem = path.getFileSystem(conf);    //根据 path 找到这个文件
//        if (fileSystem.exists(path)) {
//            fileSystem.delete(path, true);    //true 的意思是，就算 output 有东西，也一并删除
//        }

        FileInputFormat.addInputPath(job, new Path("/data/data8"));
        FileOutputFormat.setOutputPath(job, new Path("/test2/output1"));
        System.exit(job.waitForCompletion(true) ? 0 : 1);
    }

}
```

运行结果如图 7-24 所示。

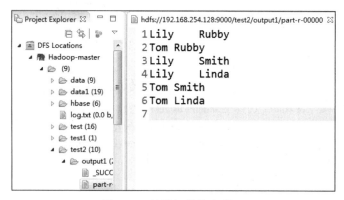

图 7-24　连接运算输出结果

7.3　MapReduce Java API 高级编程

7.3.1　多输入路径方式

多文件输入采用 7.1.1 节中提到的 MultipleInputs.addInputPath 方法即可完成，其数据集如图 7-25 和图 7-26 所示。

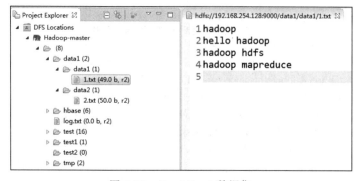

图 7-25　/data1/data1 数据集

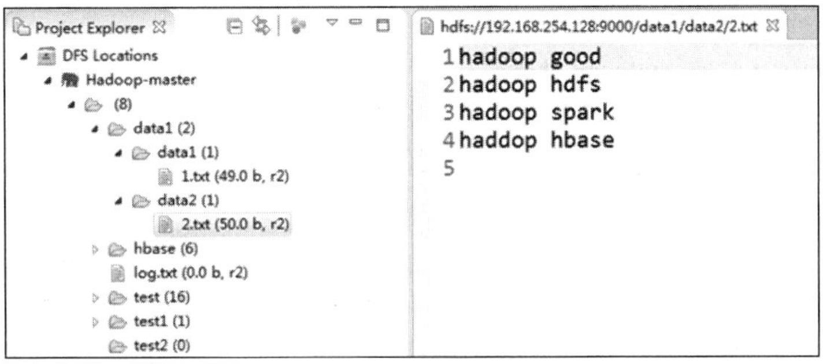

图 7-26 /data1/data2 数据集

实现代码如下：

```java
import java.io.IOException;
import java.util.StringTokenizer;
import org.apache.hadoop.conf.Configuration;
import org.apache.hadoop.fs.Path;
import org.apache.hadoop.io.IntWritable;
import org.apache.hadoop.io.LongWritable;
import org.apache.hadoop.io.Text;
import org.apache.hadoop.mapreduce.Job;
import org.apache.hadoop.mapreduce.Mapper;
import org.apache.hadoop.mapreduce.Reducer;
import org.apache.hadoop.mapreduce.lib.input.MultipleInputs;
import org.apache.hadoop.mapreduce.lib.input.SequenceFileInputFormat;
import org.apache.hadoop.mapreduce.lib.input.TextInputFormat;
import org.apache.hadoop.mapreduce.lib.output.FileOutputFormat;

public class MultiInputFormat {

    //采用 TextInputFormat
    public static class Mapper1
        extends Mapper<LongWritable, Text, Text, IntWritable>{

        private final static IntWritable one = new IntWritable(1);//1
        private Text word = new Text();

        public void map(LongWritable key, Text value, Context context
                        ) throws IOException, InterruptedException
        {
            StringTokenizer itr = new StringTokenizer(value.toString());
            while (itr.hasMoreTokens()) {
                word.set(itr.nextToken());
                //k v
                context.write(word, one);
            }
        }
    }
```

```java
public static class IntSumReducer
    extends Reducer<Text,IntWritable,Text,IntWritable> {
    private IntWritable result = new IntWritable();

    public void reduce(Text key, Iterable<IntWritable> values,
                       Context context) throws IOException, InterruptedException {
        int sum = 0;
        for (IntWritable val : values) {
            sum += val.get();
        }
        result.set(sum);
        context.write(key, result);
    }
}

public static void main(String[] args) throws Exception {
    //配置
    Configuration conf = new Configuration();
    conf.set("fs.defaultFS", "hdfs://192.168.254.128:9000");
    System.setProperty("HADOOP_USER_NAME", "root");
    //实例化一个作业，word count 是作业的名字
    Job job = Job.getInstance(conf, "word count");

    //打包运行必须执行的方法
    job.setJarByClass(MultiInputFormat.class);

    //多个输入路径
    Path path1 = new Path("/data1/data1/1.txt");
    Path path2 = new Path("/data1/data2/2.txt");
    MultipleInputs.addInputPath(job, path1, TextInputFormat.class, Mapper1.class);
    MultipleInputs.addInputPath(job, path2, TextInputFormat.class, Mapper1.class);

    //Combiner，该步骤可以省略
    job.setCombinerClass(IntSumReducer.class);

    //Reducer
    job.setReducerClass(IntSumReducer.class);
    //为 job 的输出数据设置 Key 类
    job.setOutputKeyClass(Text.class);
    //为 job 的输出数据设置 Value 类
    job.setOutputValueClass(IntWritable.class);

    //输出路径，输出路径一定是不存在的文件夹
    FileOutputFormat.setOutputPath(job, new Path("/data1/output1"));
```

```
            //提交作业
            job.waitForCompletion(true);
    }
}
```
运行结果如图 7-27 所示。

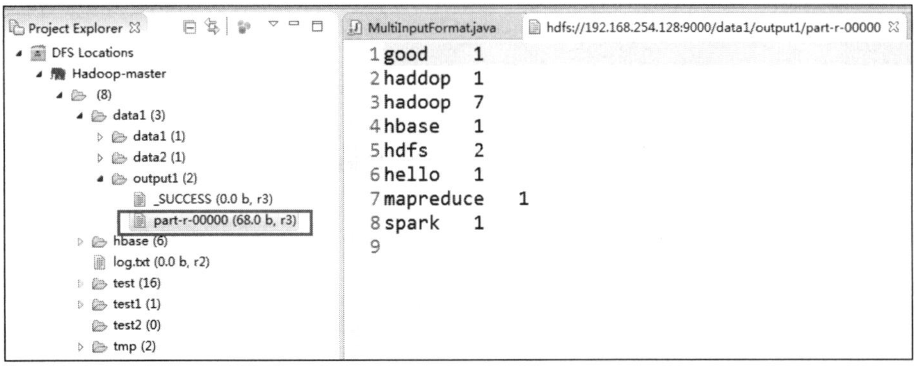

图 7-27　多输入路径统计单词出现频率

7.3.2　使用 Partitioner 实现输出到多个文件

按学生的成绩段将数据输出到不同的文件。这里分为三个成绩段：小于 60 分、大于等于 60 分小于等于 80 分和大于 80 分。Partion 发生在 Map 阶段的最后，会先调用 job.setPartitionerClass 对这个 List 进行分区，每个分区映射到一个 Reducer。每个分区内又调用 job.setSortComparatorClass 设置的 key 比较函数类排序。这里主要是用到定制 Partitioner，以下是自定义分区函数类：

public class MyPartitioner extends Partitioner<T,T>

只要继承 Partitioner<T,T>，然后去实现其中的 getPartition()方法即可，在其中完成分区的逻辑以及一些对于需求的对象 key/value 对的修改，数据集如图 7-28 所示。

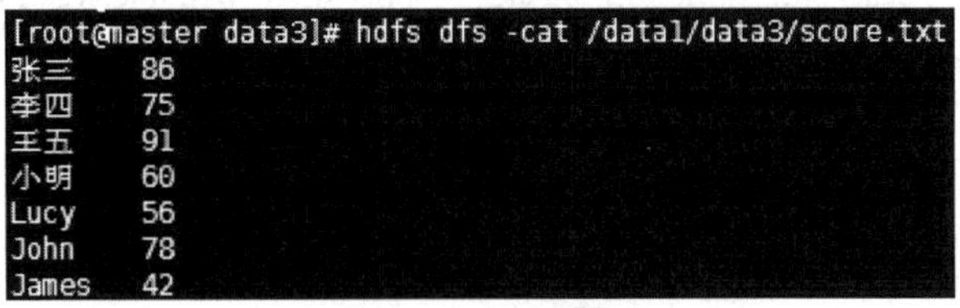

图 7-28　学生成绩数据集

实现代码如下：

```
import java.io.IOException;
import org.apache.commons.lang.StringUtils;
import org.apache.hadoop.conf.Configuration;
import org.apache.hadoop.fs.Path;
import org.apache.hadoop.io.IntWritable;
import org.apache.hadoop.io.LongWritable;
import org.apache.hadoop.io.NullWritable;
```

```java
import org.apache.hadoop.io.Text;
import org.apache.hadoop.mapreduce.Job;
import org.apache.hadoop.mapreduce.Mapper;
import org.apache.hadoop.mapreduce.Partitioner;
import org.apache.hadoop.mapreduce.Reducer;
import org.apache.hadoop.mapreduce.lib.input.FileInputFormat;
import org.apache.hadoop.mapreduce.lib.output.FileOutputFormat;

public class ScoreStats {

    public static class StudentPartitioner extends Partitioner<IntWritable, Text> {
        @Override
        public int getPartition(IntWritable key, Text value, int numReduceTasks) {
            // 学生成绩
            int scoreInt = key.get();

            // 默认指定分区 0
            if (numReduceTasks == 0)
                return 0;

            if (scoreInt < 60) { // 成绩小于 60，指定分区 0
                return 0;
            }else if (scoreInt <= 80) { // 成绩大于等于 60 小于等于 80，指定分区 1
                return 1;
            }else{ // 剩余成绩，指定分区 2
                return 2;
            }
        }
    }

    public static class StudentMapper extends Mapper<LongWritable, Text, IntWritable, Text>{
        @Override
        protected void map(LongWritable key, Text value,Context context) throws IOException, InterruptedException {
            String[] studentArr = value.toString().split("\t");

            if(StringUtils.isNotBlank(studentArr[1])){
                /*
                 * 姓名      成绩（中间以 tab 分隔）
                 * 张三       85
                 */
                // 成绩
                IntWritable pKey = new IntWritable(Integer.parseInt(studentArr[1].trim()));

                // 以成绩作为 key 输出
                context.write(pKey, value);
```

```java
            }
        }
    }

    public static class StudentReducer extends Reducer<IntWritable, Text, NullWritable, Text> {
        @Override
        protected void reduce(IntWritable key, Iterable<Text> values,Context context) throws
            IOException, InterruptedException {
            for(Text value : values){
                context.write(NullWritable.get(), value);
            }
        }
    }

    public static void main(String[] args) throws Exception {
        //读取配置文件
        //配置
        Configuration conf = new Configuration();
        conf.set("fs.defaultFS", "hdfs://192.168.254.128:9000");
        System.setProperty("HADOOP_USER_NAME", "root");

        // 新建一个任务
        Job job = Job.getInstance(conf, "Score stats");
        // 设置主类
        job.setJarByClass(StudentPartitioner.class);
        // 输入路径
        FileInputFormat.addInputPath(job, new Path("/data1/data3"));
        // 输出路径
        FileOutputFormat.setOutputPath(job, new Path("/data1/output2"));

        // Mapper
        job.setMapperClass(StudentMapper.class);
        // Reducer
        job.setReducerClass(StudentReducer.class);

        // mapper 输出格式
        job.setMapOutputKeyClass(IntWritable.class);
        job.setMapOutputValueClass(Text.class);

        // reducer 输出格式
        job.setOutputKeyClass(NullWritable.class);
        job.setOutputValueClass(Text.class);

        //设置 Partitioner 类
        job.setPartitionerClass(StudentPartitioner.class);
        // reduce 个数设置为 3
```

```
            job.setNumReduceTasks(3);

            //提交任务
            job.waitForCompletion(true);
        }
    }
```
运行结果如图 7-29 所示。

```
[root@master data3]# hdfs dfs -cat /data1/output2/part-r-00000
James    42
Lucy     56
[root@master data3]# hdfs dfs -cat /data1/output2/part-r-00001
小明     60
李四     75
John     78
[root@master data3]# hdfs dfs -cat /data1/output2/part-r-00002
张三     86
王五     91
[root@master data3]#
```

图 7-29 多文件输出结果

总结：Partitioner 适用于事先知道分区数的情况，比如像上面这个需求。但也有以下两个缺点：

（1）在作业运行之前需要知道分区数，也就是成绩段的个数，如果分区数未知，就无法操作。

（2）一般来说，让应用程序来严格限定分区数并不好，因为可能导致分区数少或者分区不均，甚至造成数据倾斜。

7.3.3 自定义 OutputFormat 文件输出

MapReduce 的输出结果默认为 part-r-00000，我们可以自定义易识别的名字来替代 part，如 score-r-00000，代码如下：

```
        job.setOutputFormatClass(MyOut.class);
        MyOut.setOutputName(job, "score");       //自定义输出名
        job.waitForCompletion(true);
        //自定义 MyOut 类继承 TextOutputFormat，并覆盖其中的 setOutputName 方法，此方法在
        //FileOutputFormat 类中为 protected 修饰，不能直接调用
        private static class MyOut extends TextOutputFormat{

            protected static void setOutputName(JobContext job, String name) {
                job.getConfiguration().set(BASE_OUTPUT_NAME, name);
            }

        }
```

上述方法仅能简单地替代文件名 part，要想全部自定义文件名，需要重写 RecordWriter。

自定义 OutputFormat 类需要继承 FileOutputFormat，并实现其中的 getRecordWriter 方法，getRecordWriter 方法返回一个 RecordWriter 对象，需要先创建此对象，实现其中的 write、close

方法。最后通过 FileSystem 在 write 方法中写出到 HDFS 自定义文件中。

自定义 OutputFormat 数据集如图 7-30 所示。

图 7-30　自定义 OutputFormat 数据集

实现代码如下：

import java.io.IOException;
import java.net.URI;
import java.net.URISyntaxException;
import java.util.StringTokenizer;
import org.apache.hadoop.conf.Configuration;
import org.apache.hadoop.fs.FSDataOutputStream;
import org.apache.hadoop.fs.FileSystem;
import org.apache.hadoop.fs.Path;
import org.apache.hadoop.io.LongWritable;
import org.apache.hadoop.io.Text;
import org.apache.hadoop.mapreduce.Job;
import org.apache.hadoop.mapreduce.JobContext;
import org.apache.hadoop.mapreduce.Mapper;
import org.apache.hadoop.mapreduce.OutputCommitter;
import org.apache.hadoop.mapreduce.OutputFormat;
import org.apache.hadoop.mapreduce.RecordWriter;
import org.apache.hadoop.mapreduce.Reducer;
import org.apache.hadoop.mapreduce.TaskAttemptContext;
import org.apache.hadoop.mapreduce.lib.input.FileInputFormat;
import org.apache.hadoop.mapreduce.lib.output.FileOutputCommitter;

public class MyOutputFormat {

  public static class MyMapper extends Mapper<LongWritable, Text, Text, LongWritable> {

    private Text word = new Text();
    private LongWritable writable = new LongWritable(1);

    @Override
    protected void map(LongWritable key, Text value, Mapper<LongWritable, Text, Text, LongWritable>.Context context)
      throws IOException, InterruptedException {
      if (value != null) {
        String line = value.toString();

```java
                StringTokenizer tokenizer = new StringTokenizer(line);
                while (tokenizer.hasMoreTokens()) {
                    word.set(tokenizer.nextToken());
                    context.write(word, writable);
                }
            }
        }
    }

    public static class MyReducer extends Reducer<Text, LongWritable, Text, LongWritable> {

        @Override
        protected void reduce(Text key, Iterable<LongWritable> values,
                Reducer<Text, LongWritable, Text, LongWritable>.Context context)
                throws IOException, InterruptedException {
            long sum = 0;
            for (LongWritable value : values) {
                sum += value.get();
            }
            context.write(key, new LongWritable(sum));
        }
    }

    public static class MyselfOutputFormat extends OutputFormat<Text, LongWritable> {

        private FSDataOutputStream outputStream = null;

        @Override
        public RecordWriter<Text, LongWritable> getRecordWriter(
                TaskAttemptContext context) throws IOException,
                InterruptedException {
            try {
                FileSystem fileSystem = FileSystem.get(new URI("/data1/output3"),
                context.getConfiguration());
                //指定文件的输出路径，不再是默认的 part-r-00000
                final Path path = new Path("/data1/output3/output");
                this.outputStream = fileSystem.create(path, false);
            } catch (URISyntaxException e) {
                e.printStackTrace();
            }
            return new MyRecordWriter(outputStream);
        }

        @Override
        public void checkOutputSpecs(JobContext context) throws IOException,
```

```java
            InterruptedException {
    }

    @Override
    public OutputCommitter getOutputCommitter(TaskAttemptContext context)
            throws IOException, InterruptedException {
        //指定_SUCCESS 输出路径
        return new FileOutputCommitter(new Path("/data1"), context);
    }

}

public static class MyRecordWriter extends RecordWriter<Text, LongWritable> {

    private FSDataOutputStream outputStream = null;

    public MyRecordWriter(FSDataOutputStream outputStream) {
        this.outputStream = outputStream;
    }

    @Override
    //指定文件输出格式
    public void write(Text key, LongWritable value) throws IOException,
            InterruptedException {
        this.outputStream.writeBytes(key.toString());
        this.outputStream.writeBytes("\t");
        this.outputStream.writeLong(value.get());
    }

    @Override
    public void close(TaskAttemptContext context) throws IOException,
            InterruptedException {
        this.outputStream.close();
    }

}

public static void main(String[] args) throws IOException, URISyntaxException,
    ClassNotFoundException, InterruptedException {
    Configuration conf = new Configuration();
    conf.set("fs.defaultFS", "hdfs://192.168.254.128:9000");
    System.setProperty("HADOOP_USER_NAME", "root");
    //删除指定输出路径
    FileSystem fileSystem = FileSystem.get(new URI("/data1/output3"), conf);
    fileSystem.delete(new Path("/data1/output3"), true);
```

```
        // 新建一个任务
        Job job = Job.getInstance(conf, "MyOutputFormat");
        // 设置主类
        job.setJarByClass(MyOutputFormat.class);
         //指定数据路径
         FileInputFormat.setInputPaths(job, new Path("/data1/data4/1.txt"));
         job.setMapperClass(MyMapper.class);
         job.setMapOutputKeyClass(Text.class);
         job.setMapOutputValueClass(LongWritable.class);

         job.setReducerClass(MyReducer.class);
         job.setOutputKeyClass(Text.class);
         job.setOutputValueClass(LongWritable.class);
         job.setOutputFormatClass(MyselfOutputFormat.class);

         job.waitForCompletion(true);
    }
}
```

自定义输出文件如图 7-31 所示。

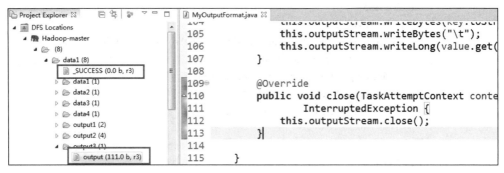

图 7-31　自定义输出文件

运行结果如图 7-32 所示。

图 7-32　自定义输出结果

总结：

（1）OutputFormat 是用于处理各种输出目的地的。

（2）OutputFormat 需要写出去的键值对来自于 Reducer 类，是通过 RecordWriter 获得的。

（3）RecordWriter 中的 write(…)方法只有 key 和 value，写到哪里去？这要通过单独传入 OutputStream 来处理，write 就是把 key 和 value 写入到 OutputStream 中的。

（4）RecordWriter 类位于 OutputFormat 中，因此我们自定义的 OutputFromat 必须继承 OutputFormat 类型。那么，流对象必须在 getRecordWriter(…)方法中获得。

7.3.4 文本文件转化成 XML 文件

通过调用 XMLOutputFormat 类（OutputFormat 的一个实现）的 getRecordWriter()方法获得 XMLRecordWriter 实例的引用。RecordWriter 配置输出的参数包括文件名的前缀和扩展名。扩展名会在代码列表中提供，文件的默认前缀为 part-r-00000。数据集如图 7-33 所示。

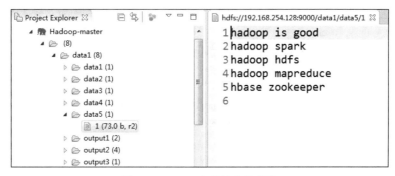

图 7-33　XML 文件转换数据集

实现代码如下：

```
import java.io.IOException;
import org.apache.hadoop.conf.Configuration;
import org.apache.hadoop.conf.Configured;
import org.apache.hadoop.fs.FSDataOutputStream;
import org.apache.hadoop.fs.FileSystem;
import org.apache.hadoop.fs.Path;
import org.apache.hadoop.io.LongWritable;
import org.apache.hadoop.io.Text;
import org.apache.hadoop.mapreduce.Job;
import org.apache.hadoop.mapreduce.Mapper;
import org.apache.hadoop.mapreduce.lib.input.FileInputFormat;
import org.apache.hadoop.mapreduce.lib.input.TextInputFormat;
import org.apache.hadoop.mapreduce.lib.output.FileOutputFormat;
import org.apache.hadoop.util.Tool;
import org.apache.hadoop.util.ToolRunner;
import java.io.DataOutputStream;
import org.apache.hadoop.mapreduce.RecordWriter;
import org.apache.hadoop.mapreduce.TaskAttemptContext;

public class XMLOutputFormatTest extends Configured implements Tool{

    public static class XMLOutputFormat extends FileOutputFormat<LongWritable,Text>{

        protected static class XMLRecordWriter extends RecordWriter<LongWritable,Text>{

            private DataOutputStream out;
```

```java
        public XMLRecordWriter(DataOutputStream out) throws IOException{
            this.out=out;
            out.writeBytes("<xml>\n");
        }
        private void writeTag(String tag,String value) throws IOException{
            if("content".equals(tag))
                out.writeBytes("<"+tag+">"+value+"</"+tag+">\n");
            else
                out.writeBytes("<"+tag+">"+value+"</"+tag+">\n\t");
        }
        @Override
        public void close(TaskAttemptContext arg0) throws IOException, InterruptedException {
            // TODO Auto-generated method stub
            try{
                out.writeBytes("</xml>\n");
            }finally{
                out.close();
            }

        }

        @Override
        public void write(LongWritable key, Text value) throws IOException, InterruptedException {
            // TODO Auto-generated method stub
            out.writeBytes("<data>\n\t");
            this.writeTag("key", Long.toString(key.get()));
            String contents[] =value.toString().split(",");
            this.writeTag("content", contents[0]);
            out.writeBytes("</data>\n");
        }
    }

    @Override
    public RecordWriter<LongWritable, Text> getRecordWriter(TaskAttemptContext job)
            throws IOException, InterruptedException {
        // TODO Auto-generated method stub
        Path file = getDefaultWorkFile(job,"");
        FileSystem fs = file.getFileSystem(job.getConfiguration());
        FSDataOutputStream fileout = fs.create(file,false);
        return new XMLRecordWriter(fileout);
    }

}
```

```java
public static class TextToXMLConversionMapper extends Mapper<LongWritable,Text,LongWritable,Text>{
    public void map(LongWritable key,Text value,Context context) throws IOException, InterruptedException{
            if(value.toString().contains(" "))
                context.write(key, value);
    }
}

@Override
public int run(String[] args) throws Exception {
    // TODO Auto-generated method stub
    Job job =Job.getInstance(getConf());
    job.setJarByClass(XMLOutputFormatTest.class);
    job.setInputFormatClass(TextInputFormat.class);
    job.setOutputFormatClass(XMLOutputFormat.class);
    job.setOutputKeyClass(LongWritable.class);
    job.setOutputValueClass(Text.class);
    job.setMapperClass(TextToXMLConversionMapper.class);
    FileInputFormat.setInputPaths(job, new Path(args[0]));
    FileOutputFormat.setOutputPath(job, new Path(args[1]));
    job.waitForCompletion(true);
    return job.isSuccessful()?0:1;
}
public static void main(String args[]) throws Exception{
    Configuration conf = new Configuration();
    conf.set("fs.defaultFS", "hdfs://192.168.254.128:9000");
    System.setProperty("HADOOP_USER_NAME", "root");
    // 新建一个任务
    Job job = Job.getInstance(conf, "XMLOutputFormat");
    job.setJarByClass(XMLOutputFormatTest.class);
    job.setJarByClass(XMLOutputFormatTest.class);
    job.setInputFormatClass(TextInputFormat.class);
    job.setOutputFormatClass(XMLOutputFormat.class);
//    job.setMapOutputKeyClass(LongWritable.class);
//    job.setMapOutputValueClass(Text.class);
    job.setOutputKeyClass(LongWritable.class);
    job.setOutputValueClass(Text.class);
    job.setMapperClass(TextToXMLConversionMapper.class);
    FileInputFormat.setInputPaths(job, new Path("/data1/data5/"));
    FileOutputFormat.setOutputPath(job, new Path("/data1/output4"));
    job.waitForCompletion(true);
}
}
```

运行结果如图 7-34 所示。

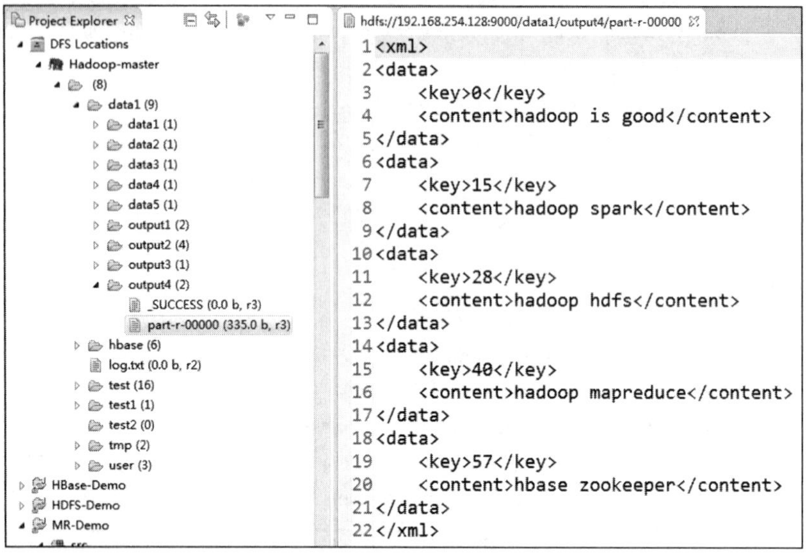

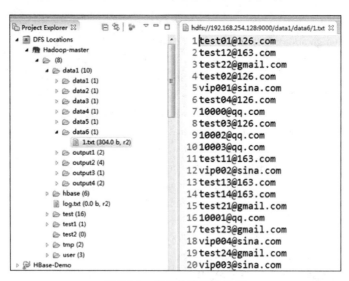

图 7-34 转换成 XML 文件输出结果

7.3.5 通过 MultipleOutputs 完成多文件输出

MultipleOutputs 类可以将数据写到多个文件中，通过该类的 write 方法即可实现多文件输出。数据集如图 7-35 所示，分别将同一域名的邮箱放入一个文件。

图 7-35 多文件输出数据集

实现代码如下：

```
import java.io.IOException;
import org.apache.hadoop.conf.Configuration;
import org.apache.hadoop.conf.Configured;
import org.apache.hadoop.fs.FileSystem;
```

```java
import org.apache.hadoop.fs.Path;
import org.apache.hadoop.io.IntWritable;
import org.apache.hadoop.io.LongWritable;
import org.apache.hadoop.io.Text;
import org.apache.hadoop.mapreduce.Job;
import org.apache.hadoop.mapreduce.Mapper;
import org.apache.hadoop.mapreduce.Reducer;
import org.apache.hadoop.mapreduce.lib.input.FileInputFormat;
import org.apache.hadoop.mapreduce.lib.output.FileOutputFormat;
import org.apache.hadoop.mapreduce.lib.output.LazyOutputFormat;
import org.apache.hadoop.mapreduce.lib.output.MultipleOutputs;
import org.apache.hadoop.mapreduce.lib.output.TextOutputFormat;

public class EmailMultipleOutputsDemo{

    public static class EmailMapper extends Mapper<LongWritable, Text, Text, IntWritable> {
        private final static IntWritable one = new IntWritable(1);

        @Override
        protected void map(LongWritable key, Text value, Context context) throws IOException, InterruptedException {
            context.write(value, one);
        }
    }

    public static class EmailReducer extends Reducer<Text, IntWritable, Text, IntWritable> {
        private MultipleOutputs<Text, IntWritable> multipleOutputs;

        @Override
        protected void setup(Context context) throws IOException ,InterruptedException{
            multipleOutputs = new MultipleOutputs< Text, IntWritable>(context);
        }

        protected void reduce(Text Key, Iterable<IntWritable> Values,Context context) throws IOException, InterruptedException {
            //开始位置
            int begin = Key.toString().indexOf("@");
            //结束位置
            int end = Key.toString().indexOf(".");

            if(begin >= end){
                return;
            }

            //获取邮箱类别，比如 qq
            String name = Key.toString().substring(begin+1, end);
```

```java
        int sum = 0;
        for (IntWritable value : Values) {
            sum += value.get();
        }

        /*
         * multipleOutputs.write(key, value, baseOutputPath)方法的第三个函数表明了该输出
           所在的目录（相对于用户指定的输出目录）。
         * 如果 baseOutputPath 不包含文件分隔符 "/"，那么输出的文件格式为
           baseOutputPath-r-nnnnn(name-r-nnnnn);
         * 如果包含文件分隔符 "/"，例如 baseOutputPath="/data1/output5/"，那么输出文件
           则为/data1/output5/part-r-nnnnn。
         */
        multipleOutputs.write(Key, new IntWritable(sum), name);
    }

    @Override
    protected void cleanup(Context context) throws IOException ,InterruptedException{
        multipleOutputs.close();
    }
}

public static void main(String[] args0) throws Exception {
    //读取配置文件
    Configuration conf = new Configuration();
    conf.set("fs.defaultFS", "hdfs://192.168.254.128:9000");
    System.setProperty("HADOOP_USER_NAME", "root");

    //判断目录是否存在，如果存在，则删除
    Path mypath = new Path("/data1/output5");
    FileSystem hdfs = mypath.getFileSystem(conf);
    if (hdfs.isDirectory(mypath)) {
        hdfs.delete(mypath, true);
    }

    //新建一个任务
    Job job = Job.getInstance(conf, "EmailMultipleOutputsDemo");
    //主类
    job.setJarByClass(EmailMultipleOutputsDemo.class);

    //输入路径
    FileInputFormat.addInputPath(job, new Path("/data1/data6/"));
    //输出路径
    FileOutputFormat.setOutputPath(job, new Path("/data1/output5"));
    // Mapper
```

```
        job.setMapperClass(EmailMapper.class);
        // Reducer
        job.setReducerClass(EmailReducer.class);
        // key 输出类型
        job.setOutputKeyClass(Text.class);
        // value 输出类型
        job.setOutputValueClass(IntWritable.class);
        //去掉 job 设置 outputFormatClass，改为通过 LazyOutputFormat 设置
        LazyOutputFormat.setOutputFormatClass(job, TextOutputFormat.class);

        job.waitForCompletion(true);
    }

}
```

运行结果如图 7-36 至图 7-40 所示。

图 7-36　126 邮箱输出统计

图 7-37　163 邮箱输出统计

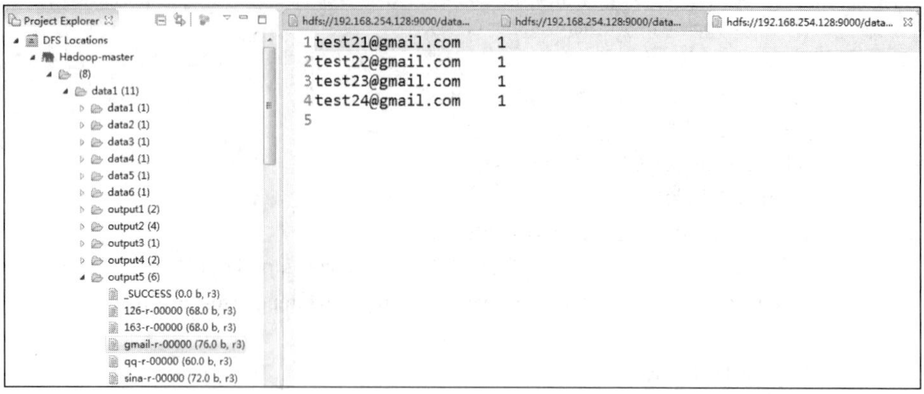

图 7-38 gmail 邮箱输出统计

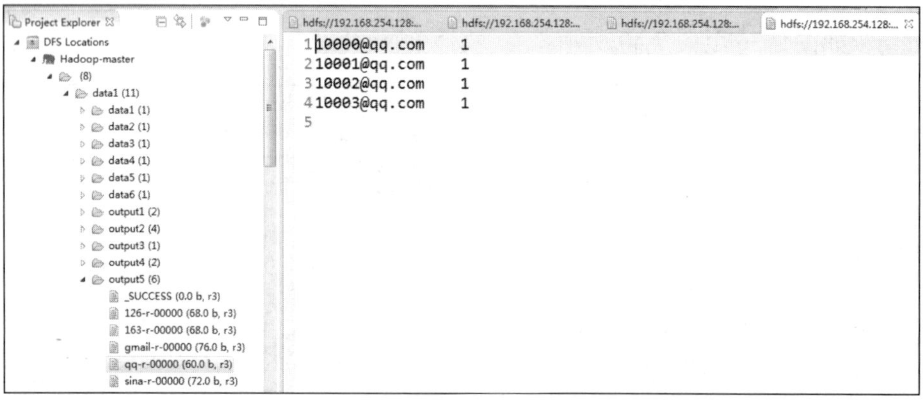

图 7-39 qq 邮箱输出统计

图 7-40 sina 邮箱输出统计

注意事项：

（1）在 Reducer 中调用时，要调用 MultipleOutputs 的接口 Public void write(KEYOUT key, VALUEOUT value, String baseOutputPath) throws IOException,InterruptedException，如果调用 public <K,V> void write(String namedOutput, K key, V value) throws IOException, InterruptedException，则需要在 Job 中预先声明 named output（代码如下），否则会报错：named output xxx not defined。

1 MultipleOutputs.addNamedOutput(job, "EmailMultipleOutputsDemo", TextOutputFormat.class, Text.class, Text.class);

2 MultipleOutputs.addNamedOutput(job, "EmailMultipleOutputsDemo", TextOutputFormat.class, Text.class, Text.class);

3 MultipleOutputs.addNamedOutput(job, "EmailMultipleOutputsDemo", TextOutputFormat.class, Text.class, Text.class);

（2）默认情况下，输出目录会生成 part-r-00000 或 part-m-00000 的空文件，需要如下设置后才不会生成：

// job.setOutputFormatClass(TextOutputFormat.class);

LazyOutputFormat.setOutputFormatClass(job, TextOutputFormat.class);

也就是去掉 job 设置 outputFormatClass，改为通过 LazyOutputFormat 设置。

（3）multipleOutputs.write(key, value, baseOutputPath)方法的第三个函数表明了该输出所在的目录（相对于用户指定的输出目录）。

7.3.6 将 MapReduce 产生的结果集导入到 MySQL 中

数据在 HDFS 和关系型数据库之间的迁移主要有以下两种方式：

（1）按照数据库要求的文件格式生成文件，然后由数据库提供的导入工具进行导入。

（2）采用 JDBC 的方式进行导入。

MapReduce 默认提供了 DBInputFormat 和 DBOutputFormat，分别用于数据库的读取和数据库的写入。下面使用 DBOutputFormat 将 MapReduce 处理后的学生信息导入到 MySQL 中。数据集如图 7-41 所示，第一列是姓名，第二列是年龄。

图 7-41 测试数据集

实现代码如下：

```
import java.io.DataInput;
import java.io.DataOutput;
import java.io.IOException;
import java.sql.PreparedStatement;
import java.sql.ResultSet;
import java.sql.SQLException;
import org.apache.commons.lang.StringUtils;
import org.apache.hadoop.conf.Configuration;
import org.apache.hadoop.conf.Configured;
import org.apache.hadoop.filecache.DistributedCache;
import org.apache.hadoop.fs.Path;
import org.apache.hadoop.io.LongWritable;
```

```java
import org.apache.hadoop.io.Text;
import org.apache.hadoop.io.Writable;
import org.apache.hadoop.mapreduce.Job;
import org.apache.hadoop.mapreduce.Mapper;
import org.apache.hadoop.mapreduce.Reducer;
import org.apache.hadoop.mapreduce.lib.db.DBConfiguration;
import org.apache.hadoop.mapreduce.lib.db.DBOutputFormat;
import org.apache.hadoop.mapreduce.lib.db.DBWritable;
import org.apache.hadoop.mapreduce.lib.input.FileInputFormat;
import org.apache.hadoop.mapreduce.lib.input.TextInputFormat;

public class MysqlDBOutputormatDemo{
    /**
     * 实现 DBWritable
     *
     * TblsWritable 需要向 MySQL 中写入数据
     */
    public static class TblsWritable implements Writable, DBWritable {
        String tbl_name;
        int tbl_age;

        public TblsWritable() {
        }

        public TblsWritable(String name, int age) {
            this.tbl_name = name;
            this.tbl_age = age;
        }

        @Override
        public void write(PreparedStatement statement) throws SQLException {
            statement.setString(1, this.tbl_name);
            statement.setInt(2, this.tbl_age);
        }

        @Override
        public void readFields(ResultSet resultSet) throws SQLException {
            this.tbl_name = resultSet.getString(1);
            this.tbl_age = resultSet.getInt(2);
        }

        @Override
        public void write(DataOutput out) throws IOException {
            out.writeUTF(this.tbl_name);
            out.writeInt(this.tbl_age);
        }
```

```java
    @Override
    public void readFields(DataInput in) throws IOException {
        this.tbl_name = in.readUTF();
        this.tbl_age = in.readInt();
    }

    public String toString() {
        return new String(this.tbl_name + " " + this.tbl_age);
    }
}

public static class StudentMapper extends Mapper<LongWritable, Text, LongWritable, Text>{
    @Override
    protected void map(LongWritable key, Text value,Context context) throws IOException,
    InterruptedException {
        context.write(key, value);
    }
}

public static class StudentReducer extends Reducer<LongWritable, Text, TblsWritable, TblsWritable> {
    @Override
    protected void reduce(LongWritable key, Iterable<Text> values,Context context) throws
    IOException, InterruptedException {
        // values 只有一个值,因为 key 没有相同的
        StringBuilder value = new StringBuilder();
        for(Text text : values){
            value.append(text);
        }

        String[] studentArr = value.toString().split("\t");

        if(StringUtils.isNotBlank(studentArr[0])){
            /*
             * 姓名    年龄 (中间以 tab 分隔)
             * 小明    16
             */
            String name = studentArr[0].trim();

            int age = 0;
            try{
                age = Integer.parseInt(studentArr[1].trim());
            }catch(NumberFormatException e){
            }

            context.write(new TblsWritable(name, age), null);
        }
    }
```

```java
        }

        public static void main(String[] args) throws Exception {
            Configuration conf = new Configuration();
            conf.set("fs.defaultFS", "hdfs://192.168.254.128:9000");
            System.setProperty("HADOOP_USER_NAME", "root");

            DBConfiguration.configureDB(conf, "com.mysql.jdbc.Driver",
                    "jdbc:mysql://192.168.254.128:3306/hadoop", "root", "123456");

            //新建一个任务
            Job job = Job.getInstance(conf, "MysqlDBOutputormatDemo");
            //设置主类
            job.setJarByClass(MysqlDBOutputormatDemo.class);

            //输入路径
            FileInputFormat.addInputPath(job, new Path("/data1/data7/student.txt"));

            //Mapper
            job.setMapperClass(StudentMapper.class);
            //Reducer
            job.setReducerClass(StudentReducer.class);

            //mapper 输出格式
            job.setOutputKeyClass(LongWritable.class);
            job.setOutputValueClass(Text.class);

            //输入格式,默认为 TextInputFormat
            job.setInputFormatClass(TextInputFormat.class);
            //输出格式
            job.setOutputFormatClass(DBOutputFormat.class);

            //输出到哪些表和字段
            DBOutputFormat.setOutput(job, "student", "name", "age");

            //添加 MySQL 数据库 jar
//            job.addArchiveToClassPath(new Path("/lib/mysql/mysql-connector-java-5.1.40-bin.jar"));
//            DistributedCache.addFileToClassPath(new Path("/lib/mysql/mysql-connector-java-5.1.40-bin.
//            jar"), conf);
            //提交任务
            job.waitForCompletion(true);
        }


}
```

MapReduce 程序很简单,只是读取文件内容,在这里我们主要关注的是怎么将 MapReduce 处理后的结果集导入 MySQL 中。数据库中表是 student,为 student 表编写对应的 Bean 类 TblsWritable,该类需要实现 Writable 接口和 DBWritable 接口。

(1) Writable 接口。

```
@Override
public void write(DataOutput out) throws IOException {
    out.writeUTF(this.tbl_name);
    out.writeInt(this.tbl_age);
}

@Override
public void readFields(DataInput in) throws IOException {
    this.tbl_name = in.readUTF();
    this.tbl_age = in.readInt();
}
```

上面两个方法对应着 Writable 接口,用于对象序列化。

(2) DBWritable 接口。

```
@Override
public void write(PreparedStatement statement) throws SQLException {
    statement.setString(1, this.tbl_name);
    statement.setInt(2, this.tbl_age);
}

@Override
public void readFields(ResultSet resultSet) throws SQLException {
    this.tbl_name = resultSet.getString(1);
    this.tbl_age = resultSet.getInt(2);
}
```

上面两个方法对应着 DBWriteable 接口。readFields 方法负责从结果集中读取数据库数据(注意 ResultSet 的下标是从 1 开始的),一次读取查询 SQL 中筛选的某一列。Write 方法负责将数据写入到数据库,将每一行的每一列依次写入。

最后进行 Job 的一些配置,具体代码如下:

```
DBConfiguration.configureDB(conf, "com.mysql.jdbc.Driver",
    "jdbc:mysql://192.168.254.128:3306/hadoop", "root", "123456");
```

上面的配置主要包括以下几项:

(1) 数据库驱动的名称:com.mysql.jdbc.Driver。
(2) 数据库 URL:jdbc:mysql:// 192.168.254.128:3306/hadoop。
(3) 用户名:root。
(4) 密码:123456。

还有以下几项需要配置:

(1) 数据库表以及每列的名称:DBOutputFormat.setOutput(job, "student", "name", "age");。
(2) 输出格式改为:job.setOutputFormatClass(DBOutputFormat.class);。

需要提醒的是,DBOutputFormat 以 MapReduce 的方式运行,会并行地连接数据库。在这里需要合理地设置 map、reduce 的个数，以便将并行连接的数量控制在合理的范围之内。

运行结果如图 7-42 所示。

```
mysql> select * from student;
+--------+------+
| name   | age  |
+--------+------+
| 小明   |   16 |
| 张三   |   18 |
| 李丹   |   25 |
| 丽丽   |   25 |
| 王五   |   36 |
| 李杨   |   31 |
| Lucy   |   20 |
| James  |   17 |
| John   |   19 |
+--------+------+
9 rows in set (0.04 sec)
```

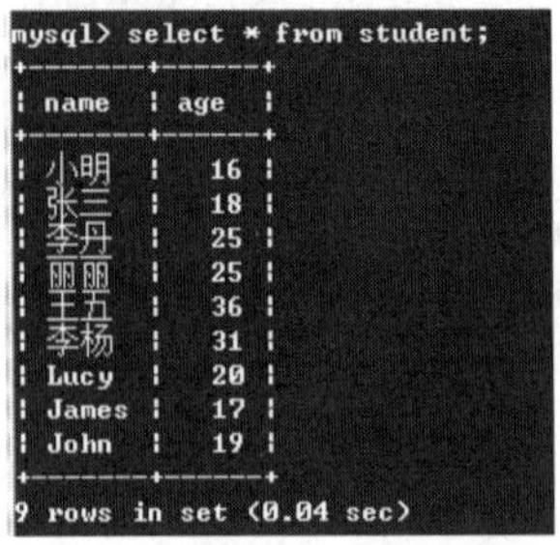

图 7-42　MySQL 数据查询输出

总结：运行程序中可能会报 com.mysql.jdbc.Driver 驱动找不到的错误，首先在本地添加所依赖的 mysql-connector-java-5.1.40-bin.jar，然后将该 JAR 包上传到 Hadoop 集群，可以通过以下方法解决：

（1）在每个节点的${HADOOP_HOME}/lib 下添加该包，重启集群，一般是比较原始的方法。

（2）把 JAR 包传到集群上，命令如下：

```
hdfs dfs -put mysql-connector-java-5.1.40-bin.jar /lib/mysql
```

在 MapReduce 程序提交 job 前添加如下语句即可：

```
job.addArchiveToClassPath(new Path("hdfs://ljc:9000/lib/mysql/mysql-connector-java-5.1.31.jar"));
```

7.3.7　自定义比较器

自定义比较器，需要继承 public static class GroupingComparator extends WritableComparator。自定义比较器，构造一个 key 对应的 value 迭代器的时候，只要 key 相同就属于同一个组，放在一个 value 迭代器中。同 key 比较函数类必须有一个构造函数，并且重载 public int compare (WritableComparable w1, WritableComparable w2)，或是实现接口 RawComparator，在 job 中设置使用 setGroupingComparatorClass。

找出每个用户的消费情况并从高到低排序，数据集如图 7-43 所示，第一列表示用户的 ID，第二列是用户名，第三列是消费金额。

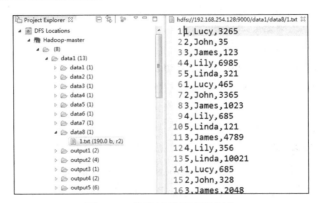

图 7-43　统计消费情况数据集

通过排序后最终会达到如下效果：

James	4789,2048,1023,123
John	3365,328,35
Lily	6985,685,356
Linda	10021,321,121
Lucy	3265,685,465

实现代码如下：

```java
import java.io.DataInput;
import java.io.DataOutput;
import java.io.IOException;
import org.apache.hadoop.io.WritableComparable;
import org.apache.hadoop.io.WritableComparator;
import org.apache.hadoop.io.IntWritable;
import org.apache.hadoop.io.LongWritable;
import org.apache.hadoop.io.Text;
import org.apache.hadoop.mapreduce.Mapper;
import org.apache.hadoop.mapreduce.Reducer;
import org.apache.hadoop.conf.Configuration;
import org.apache.hadoop.fs.FileSystem;
import org.apache.hadoop.fs.Path;
import org.apache.hadoop.mapreduce.Job;
import org.apache.hadoop.mapreduce.lib.input.FileInputFormat;
import org.apache.hadoop.mapreduce.lib.output.FileOutputFormat;

public class MyComparator {
    public static class MySort implements WritableComparable<MySort> {
        /**
         * 自定义类型中包含的变量，本例中的变量都是用于排序的变量
         * 后续的实例中我们还将定义一些其他功能的变量
         */
        private String first;
        private int second;
        public MySort() {}
        public MySort(String first, int second) {
            this.first = first;
            this.second = second;
        }
        /**
         * 反序列化，从流中的二进制转换成自定义 Key
         */
        @Override
        public void readFields(DataInput input) throws IOException {
            this.first = input.readUTF();
            this.second = input.readInt();
        }
```

```java
/**
 * 序列化，将自定义 Key 转化成使用流传送的二进制
 */
@Override
public void write(DataOutput output) throws IOException {
    output.writeUTF(first);
    output.writeInt(second);
}

public String getFirst() {
    return first;
}

public void setFirst(String first) {
    this.first = first;
}

public int getSecond() {
    return second;
}

public void setSecond(int second) {
    this.second = second;
}
/**
 * 这里不实现此方法，我们会在 SortComparator 中实现
 */
@Override
public int compareTo(MySort o) {
    return 0;
}
}

public static class SortComparator extends WritableComparator {
    public SortComparator() {
        super(MySort.class, true);
    }
    @Override
    public int compare(WritableComparable a, WritableComparable b) {
        MySort a1 = (MySort)a;
        MySort b1 = (MySort)b;
        /**
         * 首先根据第一个字段排序，然后根据第二个字段排序
         */
        if(!a1.getFirst().equals(b1.getFirst())) {
```

```java
                return a1.getFirst().compareTo(b1.getFirst());
            } else {
                return -(a1.getSecond() - b1.getSecond());
            }
        }
    }

    public static class GroupingComparator extends WritableComparator {
        public GroupingComparator() {
            super(MySort.class, true);
        }
        @Override
        public int compare(WritableComparable a, WritableComparable b) {
            MySort a1 = (MySort) a;
            MySort b1 = (MySort) b;
            /**
             * 只根据第一个字段进行分组
             */
            return a1.getFirst().compareTo(b1.getFirst());
        }
    }

    public static class MyMapper extends Mapper<LongWritable, Text, MySort, IntWritable> {
        private IntWritable cost = new IntWritable();
        @Override
        protected void map(LongWritable key, Text value, Context context)
                throws IOException, InterruptedException {
            String line = value.toString().trim();
            if(line.length() > 0) {
                String[] arr = line.split(",");
                if(arr.length == 3) {
                    cost.set(Integer.valueOf(arr[2]));
                    context.write(new MySort(arr[1],Integer.valueOf(arr[2])), cost);
                }
            }
        }
    }

    public static class MyReducer extends Reducer<MySort, IntWritable, Text, Text> {

        private Text okey = new Text();
        private Text ovalue = new Text();
        @Override
        protected void reduce(MySort key, Iterable<IntWritable> values, Context context)
                throws IOException, InterruptedException {
```

```java
            StringBuffer sb = new StringBuffer();
            /**
             * 把同一用户的消费情况进行拼接
             */
            for ( IntWritable value : values   ) {
                sb.append(",");
                sb.append(value.get());
            }
            //删除第一个逗号
            sb.delete(0, 1);
            okey.set(key.getFirst());
            ovalue.set(sb.toString());
            context.write(okey, ovalue);
        }
    }

    public static void main(String[] args) throws Exception{
        Configuration conf = new Configuration();
        conf.set("fs.defaultFS", "hdfs://192.168.254.128:9000");
        System.setProperty("HADOOP_USER_NAME", "root");
        Job job = Job.getInstance(conf, "MyComparator");
        job.setJarByClass(MyComparator.class);
        job.setMapperClass(MyMapper.class);
        job.setMapOutputKeyClass(MySort.class);
        job.setMapOutputValueClass(IntWritable.class);
        //设置排序比较器
        job.setSortComparatorClass(SortComparator.class);
        //设置分组比较器
        job.setGroupingComparatorClass(GroupingComparator.class);
        job.setReducerClass(MyReducer.class);
        job.setOutputKeyClass(Text.class);
        job.setOutputValueClass(Text.class);
        FileInputFormat.addInputPath(job, new Path("/data1/data8"));
        Path outputDir = new Path("/data1/output6");
        FileSystem fs = FileSystem.get(conf);
        if(fs.exists(outputDir)) {
            fs.delete(outputDir, true);
        }
        FileOutputFormat.setOutputPath(job, outputDir);
        job.waitForCompletion(true);
    }
}
```

运行结果如图 7-44 所示。

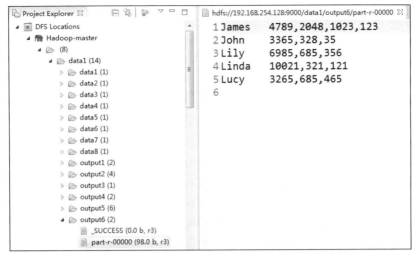

图 7-44 自定义比较器输出结果

7.3.8 MapReduce 分析明星微博数据

通过自定义输入格式，将明星微博数据排序后按粉丝数、关注数和微博数分别输出到不同文件中。数据集如图 7-45 所示。

图 7-45 明星粉丝数据集

自定义 InputFormat 读取明星微博数据，通过自定义 getSortedHashtableByValue 方法分别对明星的 fan、followers 和 microblogs 数据进行排序，然后利用 MultipleOutputs 输出不同项到不同的文件中。

实现代码如下：

```
import java.io.DataInput;
import java.io.DataOutput;
import java.io.IOException;
import org.apache.hadoop.io.WritableComparable;
import org.apache.hadoop.conf.Configuration;
import org.apache.hadoop.fs.FSDataInputStream;
import org.apache.hadoop.fs.FileSystem;
import org.apache.hadoop.fs.Path;
import org.apache.hadoop.io.Text;
import org.apache.hadoop.mapreduce.InputSplit;
import org.apache.hadoop.mapreduce.RecordReader;
import org.apache.hadoop.mapreduce.TaskAttemptContext;
import org.apache.hadoop.mapreduce.lib.input.FileInputFormat;
import org.apache.hadoop.mapreduce.lib.input.FileSplit;
```

```java
import org.apache.hadoop.util.LineReader;
import java.util.Arrays;
import java.util.Comparator;
import java.util.HashMap;
import java.util.Map;
import java.util.Map.Entry;
import org.apache.hadoop.io.IntWritable;
import org.apache.hadoop.mapreduce.Job;
import org.apache.hadoop.mapreduce.Mapper;
import org.apache.hadoop.mapreduce.Reducer;
import org.apache.hadoop.mapreduce.lib.output.FileOutputFormat;
import org.apache.hadoop.mapreduce.lib.output.LazyOutputFormat;
import org.apache.hadoop.mapreduce.lib.output.MultipleOutputs;
import org.apache.hadoop.mapreduce.lib.output.TextOutputFormat;

public class WeiBoDemo {
    //tab 分隔符
    private static String TAB_SEPARATOR = "\t";
    //粉丝
    private static String FAN = "fan";
    //关注
    private static String FOLLOWERS = "followers";
    //微博数
    private static String MICROBLOGS = "microblogs";

    public static class WeiBoMapper extends Mapper<Text, WeiBo, Text, Text> {
        @Override
        protected void map(Text key, WeiBo value, Context context) throws IOException,
        InterruptedException {
            //粉丝
            context.write(new Text(FAN), new Text(key.toString() + TAB_SEPARATOR + value.getFan()));
            //关注
            context.write(new Text(FOLLOWERS), new Text(key.toString() + TAB_SEPARATOR +
            value.getFollowers()));
            //微博数
            context.write(new Text(MICROBLOGS), new Text(key.toString() + TAB_SEPARATOR +
            value.getMicroblogs()));
        }
    }

    public static class WeiBoReducer extends Reducer<Text, Text, Text, IntWritable> {
        private MultipleOutputs<Text, IntWritable> mos;

        protected void setup(Context context) throws IOException, InterruptedException {
            mos = new MultipleOutputs<Text, IntWritable>(context);
        }
```

```java
protected void reduce(Text Key, Iterable<Text> Values,Context context) throws IOException,
InterruptedException {
    Map<String,Integer> map = new HashMap< String,Integer>();

    for(Text value : Values){
        // value =  名称 +(粉丝数或关注数或微博数)
        String[] records = value.toString().split(TAB_SEPARATOR);
        map.put(records[0], Integer.parseInt(records[1].toString()));
    }

    //对 Map 内的数据进行排序
    Map.Entry<String, Integer>[] entries = getSortedHashtableByValue(map);

    for(int i = 0; i < entries.length;i++){
        mos.write(Key.toString(),entries[i].getKey(), entries[i].getValue());
    }
}

protected void cleanup(Context context) throws IOException, InterruptedException {
    mos.close();
}
}

public static Entry<String, Integer>[] getSortedHashtableByValue(Map<String, Integer> h) {
    Entry<String, Integer>[] entries = (Entry<String, Integer>[]) h.entrySet().toArray(new Entry[0]);
    //排序
    Arrays.sort(entries, new Comparator<Entry<String, Integer>>() {
        public int compare(Entry<String, Integer> entry1, Entry<String, Integer> entry2) {
            return entry2.getValue().compareTo(entry1.getValue());
        }
    });
    return entries;
}

public static class WeiBo implements WritableComparable<Object> {
    //粉丝
    private int fan;
    //关注
    private int followers;
    //微博数
    private int microblogs;

    public WeiBo(){};

    public WeiBo(int fan,int followers,int microblogs){
        this.fan = fan;
        this.followers = followers;
```

```java
            this.microblogs = microblogs;
        }

        public void set(int fan,int followers,int microblogs){
            this.fan = fan;
            this.followers = followers;
            this.microblogs = microblogs;
        }

        //实现 WritableComparable 的 readFields()方法，以便该数据能被序列化后完成网络传输或
        //文件输入
        @Override
        public void readFields(DataInput in) throws IOException {
            fan      = in.readInt();
            followers = in.readInt();
            microblogs = in.readInt();
        }

        //实现 WritableComparable 的 write()方法，以便该数据能被序列化后完成网络传输或文件
        //输出
        @Override
        public void write(DataOutput out) throws IOException {
            out.writeInt(fan);
            out.writeInt(followers);
            out.writeInt(microblogs);
        }

        @Override
        public int compareTo(Object o) {
            // TODO Auto-generated method stub
            return 0;
        }

        public int getFan() {
            return fan;
        }

        public void setFan(int fan) {
            this.fan = fan;
        }

        public int getFollowers() {
            return followers;
        }

        public void setFollowers(int followers) {
            this.followers = followers;
```

```java
    }

    public int getMicroblogs() {
        return microblogs;
    }

    public void setMicroblogs(int microblogs) {
        this.microblogs = microblogs;
    }
}


public static class WeiboInputFormat extends FileInputFormat<Text,WeiBo>{

    @Override
    public RecordReader<Text, WeiBo> createRecordReader(InputSplit arg0,
    TaskAttemptContext arg1) throws IOException, InterruptedException {
        //自定义 WeiboRecordReader 类，按行读取
        return new WeiboRecordReader();
    }

    public class WeiboRecordReader extends RecordReader<Text, WeiBo>{
        public LineReader in;
        //声明 key 类型
        public Text lineKey = new Text();
        //声明 value 类型
        public WeiBo lineValue = new WeiBo();

        @Override
        public void initialize(InputSplit input, TaskAttemptContext context) throws IOException,
        InterruptedException {
            //获取 split
            FileSplit split = (FileSplit)input;
            //获取配置
            Configuration job = context.getConfiguration();
            //分片路径
            Path file = split.getPath();

            FileSystem fs = file.getFileSystem(job);
            //打开文件
            FSDataInputStream filein = fs.open(file);

            in = new LineReader(filein,job);
        }

        @Override
        public boolean nextKeyValue() throws IOException, InterruptedException {
```

```java
        //一行数据
        Text line = new Text();

        int linesize = in.readLine(line);

        if(linesize == 0)
            return false;

        //通过分隔符"\t"将每行的数据解析成数组
        String[] pieces = line.toString().split("\t");

        if(pieces.length != 5){
            throw new IOException("Invalid record received");
        }

        int a,b,c;
        try{
            //粉丝
            a = Integer.parseInt(pieces[2].trim());
            //关注
            b = Integer.parseInt(pieces[3].trim());
            //微博数
            c = Integer.parseInt(pieces[4].trim());
        }catch(NumberFormatException nfe){
            throw new IOException("Error parsing floating poing value in record");
        }

        //自定义key和value值
        lineKey.set(pieces[0]);
        lineValue.set(a, b, c);

        return true;
    }

    @Override
    public void close() throws IOException {
        if(in != null){
            in.close();
        }
    }

    @Override
    public Text getCurrentKey() throws IOException, InterruptedException {
        return lineKey;
    }
```

```java
        @Override
        public WeiBo getCurrentValue() throws IOException, InterruptedException {
            return lineValue;
        }

        @Override
        public float getProgress() throws IOException, InterruptedException {
            return 0;
        }

    }
}

    public static void main(String[] args) throws Exception {
        //配置文件对象
        Configuration conf = new Configuration();
        conf.set("fs.defaultFS", "hdfs://192.168.254.128:9000");
        System.setProperty("HADOOP_USER_NAME", "root");

        //判断路径是否存在，如果存在，则删除
        Path mypath = new Path("/data1/output8");
        FileSystem hdfs = mypath.getFileSystem(conf);
        if (hdfs.isDirectory(mypath)) {
            hdfs.delete(mypath, true);
        }

        //构造任务
        Job job = Job.getInstance(conf, "WeiBoDemo");
        //主类
        job.setJarByClass(WeiBoDemo.class);

        //Mapper
        job.setMapperClass(WeiBoMapper.class);
        //Mapper key 输出类型
        job.setMapOutputKeyClass(Text.class);
        //Mapper value 输出类型
        job.setMapOutputValueClass(Text.class);

        //Reducer
        job.setReducerClass(WeiBoReducer.class);
        //Reducer key 输出类型
        job.setOutputKeyClass(Text.class);
        //Reducer value 输出类型
        job.setOutputValueClass(IntWritable.class);
```

```java
    //输入路径
    FileInputFormat.addInputPath(job, new Path("/data1/data10/1.txt"));
    //输出路径
    FileOutputFormat.setOutputPath(job, new Path("/data1/output8"));

    //自定义输入格式
    job.setInputFormatClass(WeiboInputFormat.class) ;
    //自定义文件输出类别
    MultipleOutputs.addNamedOutput(job, FAN, TextOutputFormat.class, Text.class, IntWritable.class);
    MultipleOutputs.addNamedOutput(job, FOLLOWERS, TextOutputFormat.class, Text.class, IntWritable.class);
    MultipleOutputs.addNamedOutput(job, MICROBLOGS, TextOutputFormat.class, Text.class, IntWritable.class);

    //去掉 job 设置 outputFormatClass，改为通过 LazyOutputFormat 设置
    LazyOutputFormat.setOutputFormatClass(job, TextOutputFormat.class);

     //提交任务
    job.waitForCompletion(true);
    }

}
```

运行结果如图 7-46 所示。

```
[root@master data]# hdfs dfs -cat /data1/output8/fan-r-00000
何炅      90165654
杨幂      78387383
范冰冰    60963113
赵丽颖    58407341
王俊凯    35684830
佟丽娅    30617838
[root@master data]# hdfs dfs -cat /data1/output8/followers-r-00000
佟丽娅    720
何炅      714
杨幂      577
赵丽颖    457
范冰冰    248
王俊凯    227
[root@master data]# hdfs dfs -cat /data1/output8/microblogs-r-00000
何炅      8259
佟丽娅    4970
杨幂      3623
范冰冰    1794
赵丽颖    1774
王俊凯    929
```

图 7-46 明细粉丝统计信息输出

7.3.9 MapReduce 最佳成绩统计

统计出 0~20、20~50、50~100 这三个年龄段的男女学生的最高分数。数据集如图 7-47 所示，第一列是姓名，第二列是年龄，第三列是性别，第四列是成绩。

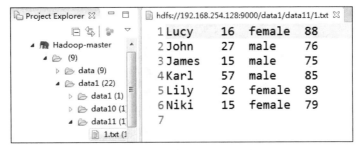

图 7-47 男女生成绩数据集

（1）编写 Mapper 类，按需求将数据集解析为 key=gender，value=name+age+score，然后输出。

（2）编写 Partitioner 类，按年龄段将结果指定给不同的 Reduce 执行。

（3）编写 Reducer 类，分别统计出男女学生的最高分数。

实现代码如下：

```
import java.io.IOException;
import org.apache.hadoop.conf.Configuration;
import org.apache.hadoop.fs.FileSystem;
import org.apache.hadoop.fs.Path;
import org.apache.hadoop.io.LongWritable;
import org.apache.hadoop.io.Text;
import org.apache.hadoop.mapreduce.Job;
import org.apache.hadoop.mapreduce.Mapper;
import org.apache.hadoop.mapreduce.Partitioner;
import org.apache.hadoop.mapreduce.Reducer;
import org.apache.hadoop.mapreduce.lib.input.FileInputFormat;
import org.apache.hadoop.mapreduce.lib.output.FileOutputFormat;


public class GenderSortDemo{
    private static String TAB_SEPARATOR = "\t";

    public static class GenderMapper extends Mapper<LongWritable, Text, Text, Text> {
        /*
         * 调用 map 解析一行数据，该行数据存储在 value 参数中，然后根据"\t"分隔符解析出
         * 姓名、年龄、性别和成绩
         */
        public void map(LongWritable key, Text value, Context context) throws IOException,
        InterruptedException {
            /*
             * 姓名      年龄       性别        成绩
             * Lucy      23        female      65
             * 每个字段的分隔符是 tab 键
             */
            //使用"\t"分隔数据
            String[] tokens = value.toString().split(TAB_SEPARATOR);
```

```java
            //性别
            String gender = tokens[2];
            //姓名    年龄    成绩
            String nameAgeScore = tokens[0] + TAB_SEPARATOR + tokens[1] + TAB_SEPARATOR +
            tokens[3];

            //输出   key=gender    value=name+age+score
            context.write(new Text(gender), new Text(nameAgeScore));
        }
    }

    /*
     * 合并 Mapper 输出结果
     */
    public static class GenderCombiner extends Reducer<Text, Text, Text, Text>   {

        public void reduce(Text key, Iterable<Text> values, Context context)throws IOException,
        InterruptedException {
            int maxScore = Integer.MIN_VALUE;
            int score = 0;
            String name = " ";
            String age = " ";

            for (Text val : values) {
                String[] valTokens = val.toString().split(TAB_SEPARATOR);
                score = Integer.parseInt(valTokens[2]);
                if (score > maxScore) {
                    name = valTokens[0];
                    age = valTokens[1];
                    maxScore = score;
                }
            }

            context.write(key, new Text(name + TAB_SEPARATOR + age + TAB_SEPARATOR +
            maxScore));
        }
    }

    /*
     * 根据 age（年龄）段将 map 输出结果均匀分布在 reduce 上
     */
    public static class GenderPartitioner extends Partitioner<Text, Text> {

        @Override
        public int getPartition(Text key, Text value, int numReduceTasks) {
            String[] nameAgeScore = value.toString().split(TAB_SEPARATOR);
            //学生年龄
```

```java
        int age = Integer.parseInt(nameAgeScore[1]);

        //默认指定分区 0
        if (numReduceTasks == 0)
            return 0;

        //年龄小于等于 20,指定分区 0
        if (age <= 20) {
            return 0;
        }else if (age <= 50) { //年龄大于 20 小于等于 50,指定分区 1
            return 1 % numReduceTasks;
        }else         //剩余年龄,指定分区 2
            return 2 % numReduceTasks;
    }
}

/*
 * 统计出不同性别的最高分数
 */
    public static class GenderReducer extends Reducer<Text, Text, Text, Text>  {
        @Override
        public void reduce(Text key, Iterable<Text> values, Context context) throws IOException,
InterruptedException {
            int maxScore = Integer.MIN_VALUE;
            int score = 0;
            String name = " ";
            String age = " ";
            String gender = " ";

            //根据 key 迭代 values 集合,求出最高分数
            for (Text val : values) {
                String[] valTokens = val.toString().split(TAB_SEPARATOR);
                score = Integer.parseInt(valTokens[2]);
                if (score > maxScore) {
                    name = valTokens[0];
                    age = valTokens[1];
                    gender = key.toString();
                    maxScore = score;
                }
            }

            context.write(new Text(name), new Text("age: " + age + TAB_SEPARATOR + "gender: "
+ gender + TAB_SEPARATOR + "score: " + maxScore));
        }
    }


    public static void main(String[] args) throws Exception {
```

```java
//读取配置文件
Configuration conf = new Configuration();
conf.set("fs.defaultFS", "hdfs://192.168.254.128:9000");
System.setProperty("HADOOP_USER_NAME", "root");

Path mypath = new Path("/data1/output9");
FileSystem hdfs = mypath.getFileSystem(conf);
if (hdfs.isDirectory(mypath)) {
    hdfs.delete(mypath, true);
}

//新建一个任务
Job job = Job.getInstance(conf, "GenderSortDemo");
//主类
job.setJarByClass(GenderSortDemo.class);
//Mapper
job.setMapperClass(GenderMapper.class);
//Reducer
job.setReducerClass(GenderReducer.class);

//map 输出 key 类型
job.setMapOutputKeyClass(Text.class);
//map 输出 value 类型
job.setMapOutputValueClass(Text.class);

//reduce 输出 key 类型
job.setOutputKeyClass(Text.class);
//reduce 输出 value 类型
job.setOutputValueClass(Text.class);

//设置 Combiner 类
job.setCombinerClass(GenderCombiner.class);

//设置 Partitioner 类
job.setPartitionerClass(GenderPartitioner.class);
//reduce 个数设置为 3
job.setNumReduceTasks(3);

//输入路径
FileInputFormat.addInputPath(job, new Path("/data1/data11/"));
//输出路径
FileOutputFormat.setOutputPath(job, new Path("/data1/output9"));

//提交任务
job.waitForCompletion(true);
    }
}
```

运行结果如图 7-48 至图 7-50 所示。

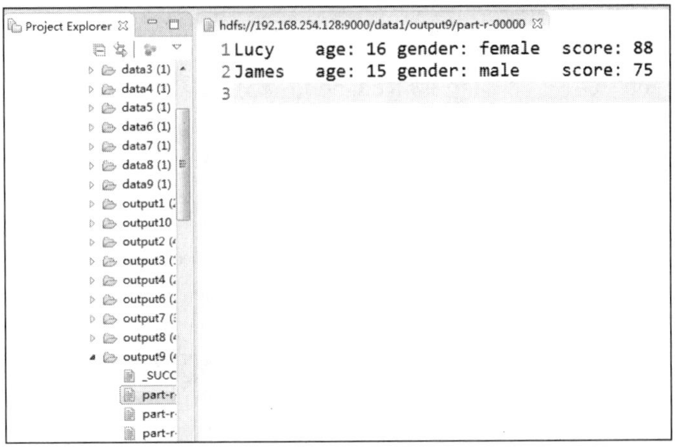

图 7-48　0～20 年龄段男女最高成绩输出

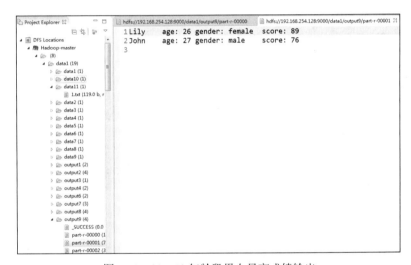

图 7-49　20～50 年龄段男女最高成绩输出

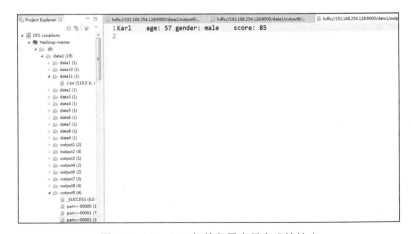

图 7-50　50～100 年龄段男女最高成绩输出

7.3.10 MapReduce 链接作业

对于简单的程序，我们只需要一个 MapReduce 就能搞定，然而对于比较复杂的程序，我们可能需要多个 Job 或者多个 Map 或 Reduce 进行计算。下面就来介绍多个 Job 或者多个 MapReduce 的编程形式。

1. 迭代式 MapReduce

MapReduce 迭代方式，通常是将上一个 MapReduce 任务的输出作为下一个 MapReduce 任务的输入，可以只保留 MapReduce 任务的最终结果，中间数据可以删除或保留，如图 7-51 所示。

图 7-51　迭代方式 MapReduce 运行图

迭代式 MapReduce 的示例代码如下：

```
public class IterativeJob{
    //这里只给出主要代码，其他省略
    ...

    @Override
    public int run(String[] args) throws Exception {
        Configuration conf = new Configuration();

        //第一个 MapReduce 任务
        Job job1 = new Job(conf,"job1");
        ...
        //job1 的输入
        FileInputFormat.addInputPath(job1,input);
        //job1 的输出
        FileOutputFromat.setOutputPath(job1,out1);
        job1.waitForCompletion(true);

        //第二个 Mapreduce 任务
        Job job2 = new Job(conf,"job2");
        ...
        //job1 的输出作为 job2 的输入
        FileInputFormat.addInputPath(job2,out1);
        //job2 的输出
        FileOutputFromat.setOutputPath(job2,out2);
        job2.waitForCompletion(true);

        //第三个 Mapreduce 任务
        Job job3 = new Job(conf,"job3");
        ...
```

```
        //job2 的输出作为 job3 的输入
        FileInputFormat.addInputPath(job3,out2);
        //job3 的输出
        FileOutputFromat.setOutputPath(job3,out3);
        job3.waitForCompletion(true);
        ...
    }

        ...
}
```

虽然 MapReduce 的迭代可以实现多任务的执行，但是它具有以下两个缺点：

（1）每次迭代，如果所有 Job 对象重复创建，代价将非常高。

（2）每次迭代，数据都要写入本地，然后从本地读取，I/O 和网络传输的代价比较大。

2．依赖式 MapReduce

依赖式 MapReduce 是由 org.apache.hadoop.mapred.jobcontrol 包中的 JobControl 类来实现的。JobControl 的实例表示一个作业的运行图，可以加入作业配置，然后告知 JobControl 实例作业之间的依赖关系。在一个线程中运行 JobControl 时，它将按照依赖顺序来执行这些作业。也可以查看进程，在作业结束后，可以查询作业的所有状态和每个失败相关的错误信息。如果一个作业失败，JobControl 将不执行与之有依赖关系的后续作业。依赖式 MapReduce 运行图如图 7-52 所示。

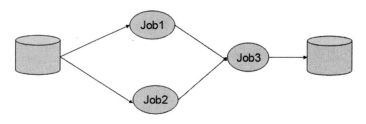

图 7-52　依赖式 MapReduce 运行图

```
public class DependentJob{
    //这里只给出主要代码，其他省略
    ...

    @Override
    public int run(String[] args) throws Exception {
        Configuration conf1 = new Configuration();
        Job job1 = new Job(conf1,"Job1");
        ...

        Configuration conf2 = new Configuration();
        Job job2 = new Job(conf2,"Job2");
        ...

        Configuration conf3 = new Configuration();
```

```
Job job3 = new Job(conf3,"Job3");
...

//构造一个 ControlledJob
ControlledJob cJob1 = new ControlledJob(conf1);
//设置 ControlledJob
cJob1.setJob(job1);
ControlledJob cJob2 = new ControlledJob(conf2);
cJob2.setJob(job2);
ControlledJob cJob3 = new ControlledJob(conf3);
cJob2.setJob(job3);

//设置 cJob3 和 cJob1 的依赖关系
cJob3.addDependingJob(cJob1);
//设置 cJob3 和 cJob2 的依赖关系
cJob3.addDependingJob(cJob2);

JobControl JC = new JobControl("dependentJob");
//把 3 个构造的 ControlledJob 加入到 JobControl 中
JC.addJob(cJob1);
JC.addJob(cJob2);
JC.addJob(cJob3);
Thread t = new Thread(JC);
t.start();
while (true) {
    if (jobControl.allFinished()) {
        jobControl.stop();
        break;
    }
}
}
...
}
```

注意：Hadoop 的 JobControl 类实现了线程 Runnable 接口。我们需要实例化一个线程来启动它。直接调用 JobControl 的 run()方法，线程将无法结束。

3. 链式 MapReduce

大量的数据处理任务涉及对记录的预处理和后处理。例如，在处理信息检索的文档时，可能需要先移除像 a、the 和 is 这样经常出现但不太有意义的词，然后再作转换，将一个词的不同形式转为相同的形式，例如将 finishing 和 finished 转换为 finish。

我们可以为预处理与后处理各自编写一个 MapReduce 作业，并把它们链接起来。在这些作业中可以使用 IdentityReducer（或完全不同的 Reducer）。由于在执行过程中每一个作业的中间结果都需要占用 I/O 和存储资源，所以这种做法是低效的。另一种方法是自己写 Mapper 去预先调用所有的预处理作业，再让 Reducer 调用所有的后处理作业。这将强制我们采用模块化和可组合的方式来构建预处理和后处理，因此 Hadoop 引入了 ChainMapper 和 ChainReducer 类来简化预处理和后处理的构成。

Hadoop 提供了专门的链式 ChainMapper 和 ChainReducer 来处理链式 MapReduce 任务。在 Map 或者 Reduce 阶段存在多个 Mapper，这些 Mapper 像 Linux 管道一样，前一个 Mapper 的输出结果直接重定向到后一个 Mapper 的输入，形成流水线，如图 7-53 所示。

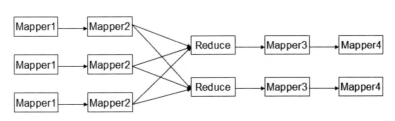

图 7-53　链式 MapReduce 运行图

其调用形式如下：

　　...
　　ChainMapper.addMapper(...);
　　ChainReducer.setReducer(...);
　　ChainReducer.addMapper(...);
　　...

addMapper 方法如下：

　　public static void addMapper(Job job,
　　　　Class<extends Mapper> mclass,
　　　　Class<extends K1> inputKeyClass,
　　　　Class<extends V1> inputValueClass,
　　　　Class<extends K2> outputKeyClass,
　　　　Class<extends V2> outputValueClass,
　　　　Configuration conf
　　)

addMapper()方法有 7 个参数：第一个和最后一个分别为全局的 Job 和本地的 configuration 对象；第二个参数是 Mapper 类，负责数据处理；余下的 4 个参数 inputKeyClass、inputValueClass、outputKeyClass 和 outputValueClass 是这个 Mapper 类中输入/输出类的类型。ChainReducer 专门提供了一个 setReducer()方法来设置整个作业唯一的 Reducer，语法与 addMapper()方法类似。

链式 MapReduce 的示例代码如下：

```
public class ChainJob extends Configured implements Tool {
    // 这里只给出主要代码，其他省略
    ...

    @Override
    public int run(String[] args) throws Exception {
        Configuration conf = new Configuration();
        Job job = new Job(conf);
```

```
        job.setJobName("chainjob");
        job.setInputFormat(TextInputFormat.class);
        job.setOutputFormat(TextOutputFormat.class);

        FileInputFormat.addInputPath(job, in);
        FileOutputFormat.setOutputPath(job, out);

        //在作业中添加 Map1 阶段
        Configuration map1conf = new Configuration(false);
        ChainMapper.addMapper(job, Map1.class, LongWritable.class, Text.class,Text.class, Text.class,
        map1conf);

        //在作业中添加 Map2 阶段
        Configuration map2conf = new Configuration(false);
        ChainMapper.addMapper(job, Map2.class, Text.class, Text.class,LongWritable.class, Text.class,
        map2conf);

        //在作业中添加 Reduce 阶段
        Configuration reduceconf = new Configuration(false);
        ChainReducer.setReducer(job,Reduce.class,LongWritable.class,Text.class,Text.class,Text.class,
        reduceconf);

        //在作业中添加 Map3 阶段
        Configuration map3conf = new Configuration(false);
        ChainReducer.addMapper(job,Map3.class,Text.class,Text.class,LongWritable.class,Text.class,
        map3conf);

        //在作业中添加 Map4 阶段
        Configuration map4conf = new Configuration(false);
        ChainReducer.addMapper(job,Map4.class,LongWritable.class,Text.class,LongWritable.class,
        Text.class,map4conf);

        job.waitForCompletion(true);
    }

    ...
}
```

注意：对于任意一个 MapReduce 作业，Map 和 Reduce 阶段可以有无限个 Mapper，但是 Reduce 只能有一个。所以包含多个 Reduce 的作业，不能使用 ChainMapper/ChainReduce 来完成。

7.3.11　利用 Job 嵌套求解二度人脉

任务是求解其中的二度人脉、潜在好友，也就是如图 7-54 所示。

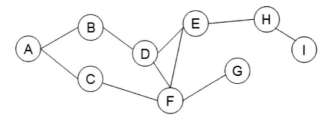

图 7-54 二度人脉

例如 A 认识 B 和 C，但 B 不认识 C，那么 B-C 就是一对潜在好友，虽然 D 认识 B、E 和 F，但 E-F 不是潜在好友，也就是两者不能直连。数据集如图 7-55 所示。

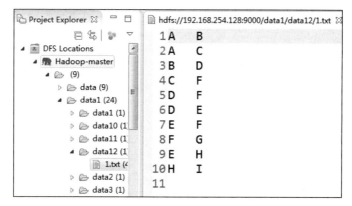

图 7-55 二度人脉数据集

首先，我们进行第一个 MapReduce。

（1）与 7.2.10 节利用单表关联在父子关系中求解爷孙关系相同，同样是一个输入行，产生一对互逆的关系，压入 context。

例如 A　B 这个输入行就在 Map 阶段整理出 A　B-B　A 这样的互逆关系。

（2）之后 Reduce 会自动将 context 中相同的 key 合并在一起。

例如，由于存在 A　B、A　C，显然会产生一个 A:{B,C}，这是 Reduce 阶段开始的键值对。

（3）这个键值对相当于 A 所认识的人。先进行如下的输出，1 代表 A 的一度人脉：

  A　　B　　1
  A　　C　　1

潜在好友显然会在{B, C}这个 A 所认识的人中产生，对这个数组做笛卡尔乘积，形成关系：{<B,B>,<C,C>,<B,C>,<C,B>}。

将存在自反性，前项首字母大于后项剔除，也就是<B,B>、<C,C>这类无意义的剔除，<B,C>、<C,B>认定为一个关系，将剩余关系进行如下的输出，其中 2 代表 B 的二度人脉，也就是 B 所谓的潜在好友：

  B　　C　　2

此时，第一个 MapReduce 输出如下：

  A　　C　　1
  A　　B　　1

B	C	2
B	D	1
A	B	1
A	D	2
C	F	1
A	C	1
A	F	2
B	D	1
D	E	1
D	F	1
B	E	2
B	F	2
E	F	2
E	H	1
E	F	1
D	E	1
F	H	2
D	H	2
D	F	2
D	F	1
F	G	1
E	F	1
C	F	1
D	G	2
D	E	2
E	G	2
C	D	2
C	G	2
C	E	2
F	G	1
E	H	1
H	I	1
E	I	2
H	I	1

这时，数据已经很明显了，再进行第二个 MapReduce，任务是剔除本身就存在的关系，也就是在潜在好友中剔除本身就认识的关系。

将上述第一个 MapReduce 的输出关系作为 key，后面的 X 度人脉这个 1、2 值作为 value，进行 MapReduce 处理。那么，例如<B,C>这个关系，之所以会被认定为潜在好友，是因为它所对应的值数组里面一个 1 都没有，全是 2，也就是它们本来不是一度人脉，而<E,F>这对不能成为潜在好友，因为他们所对应的值数组里面有 1，存在任意一对一度人脉，直接认识，就绝对不能被认定为二度人脉。

将被认定为二度人脉的关系输出，就得到最终结果。其实这个 MapReduce 就是做了一件类似去除重复行的操作，只是在 Reduce 中增加了一个输出判断。

实现代码如下:

```java
import java.io.IOException;
import java.util.ArrayList;
import java.util.Random;
import org.apache.hadoop.conf.Configuration;
import org.apache.hadoop.fs.FileSystem;
import org.apache.hadoop.fs.Path;
import org.apache.hadoop.io.Text;
import org.apache.hadoop.mapreduce.Job;
import org.apache.hadoop.mapreduce.Mapper;
import org.apache.hadoop.mapreduce.Reducer;
import org.apache.hadoop.mapreduce.lib.input.FileInputFormat;
import org.apache.hadoop.mapreduce.lib.output.FileOutputFormat;
import org.apache.hadoop.util.GenericOptionsParser;

public class ChainMRDemo {

    //第一轮 MapReduce
    public static class Job1_Mapper extends Mapper<Object, Text, Text, Text> {
        public void map(Object key, Text value, Context context)
                throws IOException, InterruptedException {
            String[] line = value.toString().split("\t");     //输入文件,键值对的分隔符为空格
            context.write(new Text(line[0]), new Text(line[1]));
            context.write(new Text(line[1]), new Text(line[0]));
        }
    }

    public static class Job1_Reducer extends Reducer<Text, Text, Text, Text> {
        public void reduce(Text key, Iterable<Text> values, Context context)
                throws IOException, InterruptedException {
            ArrayList<String> potential_friends = new ArrayList<String>();

            for (Text v : values) {
                potential_friends.add(v.toString());
                if (key.toString().compareTo(v.toString()) < 0) {//确保首字母大者在前,如 Tom
                                                                 //Alice,则输出 Alice Tom
                    context.write(new Text(key + "\t" + v), new Text("1"));
                } else {
                    context.write(new Text(v + "\t" + key), new Text("1"));
                }
            }
            for (int i = 0; i < potential_friends.size(); i++) {//潜在好友集合自己与自己做笛卡尔
                                                                //乘积,求出潜在的二度人脉关系
                for (int j = 0; j < potential_friends.size(); j++) {
```

```java
                            if (potential_friends.get(i).compareTo(//将存在自反性，前项首字母大于后
                                                                  //项的关系剔除
                                    potential_friends.get(j)) < 0) {
                                context.write(new Text(potential_friends.get(i) + "\t"
                                        + potential_friends.get(j)), new Text("2"));
                            }
                        }
                    }
                }
            }

            //第二轮 MapReduce
            public static class Job2_Mapper extends Mapper<Object, Text, Text, Text> {
                public void map(Object key, Text value, Context context)
                        throws IOException, InterruptedException {
                    String[] line = value.toString().split("\t");    //输入文件，键值对的分隔符为"\t"
                    //关系作为 key，后面的 X 度人脉这个 1、2 值作为 value
                    context.write(new Text(line[0] + "\t" + line[1]), new Text(line[2]));
                }
            }

            public static class Job2_Reducer extends Reducer<Text, Text, Text, Text> {
                public void reduce(Text key, Iterable<Text> values, Context context)
                        throws IOException, InterruptedException {
                    //检查合并之后是否存在任意一对一度人脉
                    boolean is_potential_friend = true;
                    for (Text v : values) {
                        if (v.toString().equals("1")) {
                            is_potential_friend = false;
                            break;
                        }
                    }
                    //如果没有，则输出
                    if (is_potential_friend) {
                        String[] potential_friends = key.toString().split("\t");
                        context.write(new Text(potential_friends[0]), new Text(
                                potential_friends[1]));
                    }
                }
            }

            public static void main(String[] args) throws Exception {
                Configuration conf = new Configuration();
                conf.set("fs.defaultFS", "hdfs://192.168.254.128:9000");
                System.setProperty("HADOOP_USER_NAME", "root");
```

```java
//          String[] otherArgs = new GenericOptionsParser(conf, args)
//                  .getRemainingArgs();
//          if (otherArgs.length != 2) {
//              System.err.println("Usage: wordcount <in> <out>");
//              System.exit(2);
//          }

            // 判断 output 文件夹是否存在，如果存在则删除
//          Path path = new Path(otherArgs[1]);    //取第 1 个表示输出目录参数（第 0 个参数是输入目录）
//          FileSystem fileSystem = path.getFileSystem(conf);    //根据 path 找到这个文件
//          if (fileSystem.exists(path)) {
//              fileSystem.delete(path, true);    //true 的意思是，就算 output 有东西，也一并删除
//          }

            //设置第一轮 MapReduce 的相应处理类与输入输出
            Job job1 = Job.getInstance(conf, "ChainMRDemo1");
            job1.setMapperClass(Job1_Mapper.class);
            job1.setReducerClass(Job1_Reducer.class);
            job1.setOutputKeyClass(Text.class);
            job1.setOutputValueClass(Text.class);

            //定义一个临时目录，先将任务的输出结果写到临时目录中，下一个 job 以临时目录为
            //输入目录
//          FileInputFormat.addInputPath(job1, new Path(otherArgs[0]));
            FileInputFormat.addInputPath(job1, new Path("/data1/data12"));
            Path tempDir = new Path("/data1/temp_"
                    + Integer.toString(new Random().nextInt(Integer.MAX_VALUE)));
            FileOutputFormat.setOutputPath(job1, tempDir);

            if (job1.waitForCompletion(true)) {//如果第一轮 MapReduce 完成再做这里的代码
                Job job2 = Job.getInstance(conf, "ChainMRDemo2");
                FileInputFormat.addInputPath(job2, tempDir);
                //设置第二轮 MapReduce 的相应处理类与输入输出
                job2.setMapperClass(Job2_Mapper.class);
                job2.setReducerClass(Job2_Reducer.class);
//              FileOutputFormat.setOutputPath(job2, new Path(otherArgs[1]));
                FileOutputFormat.setOutputPath(job2, new Path("/data1/output10"));
                job2.setOutputKeyClass(Text.class);
                job2.setOutputValueClass(Text.class);
                FileSystem.get(conf).deleteOnExit(tempDir);    //删除刚刚临时创建的输入目录
                System.exit(job2.waitForCompletion(true) ? 0 : 1);
            }
        }
    }

}
```

运行结果如图 7-56 所示。

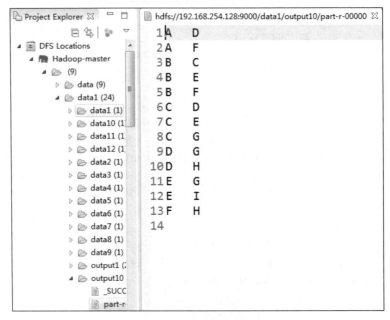

图 7-56　二度人脉结果输出

7.4　本章小结

本章详细介绍了 MapReduce 程序的具体编写方法，包括编写 InputFormat、Mapper、Combiner、Partitioner、Reducer 和 OutputFormat。在应用实例中，讲解了统计单词频率、统计平均成绩、关系运算等基本操作。在高级编程中，讲解了多文件输入/输出、自定义输出文件名、MySQL 交互操作、自定义比较器等。通过本章的学习，可以形成大家对 MapReduce 编程方法的基本认识。

附录　CentOS7 安装

CentOS7 64 位操作系统安装在虚拟机 VMware9 中，下载地址是http://isoredirect.centos.org/centos/7/isos/x86_64/CentOS-7-x86_64-DVD-1708.iso。

（1）在 VMware9 中选择 Create a New Virtual Machine，然后选择第一项安装模式，如附图 1 所示。

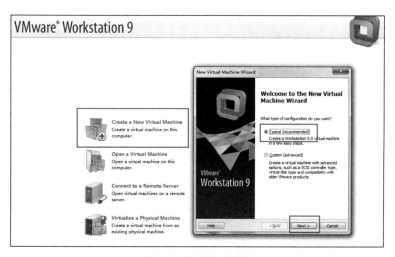

附图 1　创建新虚拟机

（2）选择操作系统安装源，这里选择第三项稍后安装，如附图 2 所示。

附图 2　指定安装源

（3）选择要安装的操作系统类型及版本号，如附图 3 所示。

（4）指定虚拟机操作系统的名字及安装目录，如附图 4 所示。

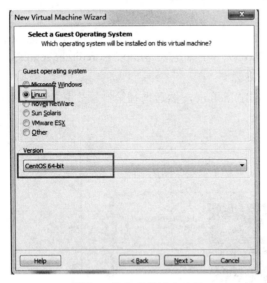 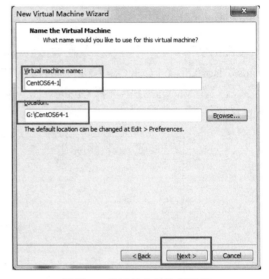

附图 3　操作系统版本选择　　　　　附图 4　虚拟机安装路径设置

（5）指定虚拟机硬盘容量（默认为 20GB），如果存放大量数据可以根据需要进行调整，如附图 5 所示。

（6）虚拟机硬件配置，将内存设置为 2GB，指定安装源路径和网络连接类型，如附图 6 至附图 9 所示。

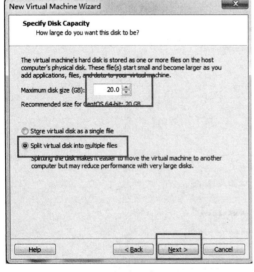 

附图 5　虚拟机硬盘大小设置　　　　附图 6　定制虚拟机硬件

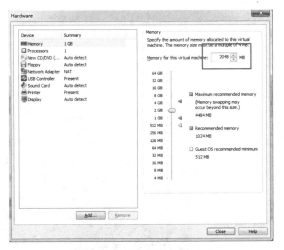

附图 7 设置虚拟机内存大小

附图 8 指定安装源

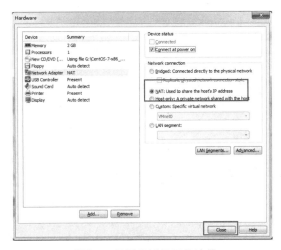

附图 9 配置虚拟机网络连接

（7）配置完成后即可看到之前配置的虚拟机名和相关硬件设置，如附图 10 所示。

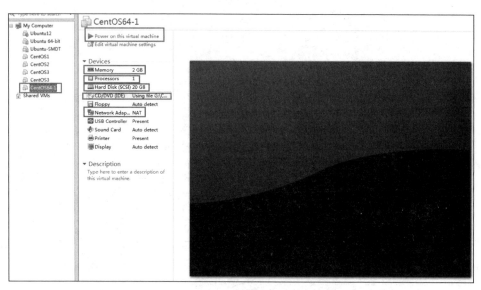

附图 10　虚拟机信息

（8）单击 Power on this virtual machine 启动该虚拟机进行 CentOS 的安装，将操作系统语言设置为中文，如附图 11 所示。

附图 11　语言选择

（9）软件选择。如果是 mini 安装，不用选择需要安装的选项；如果是桌面版安装，选择需要安装的软件（可以使用 GNOME 或 KDE，这里使用了 GNOME 桌面），如附图 12 和附图 13 所示。

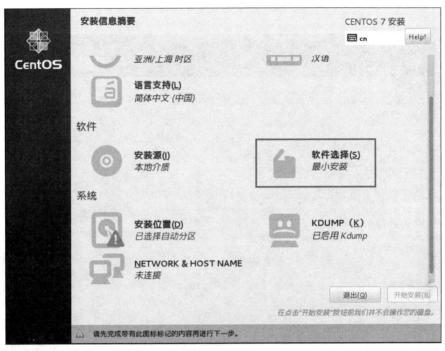

附图 12　软件选择

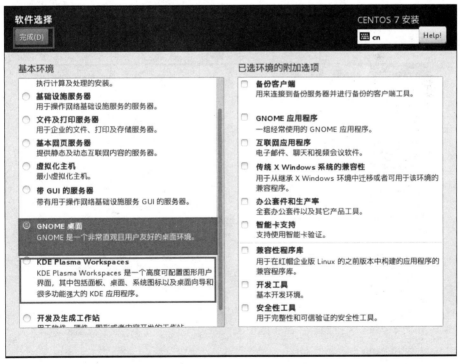

附图 13　桌面版本安装软件选择

（10）选择硬盘分区方案，可以使用自带的自动分区，也可以手动指定硬盘分区，如附图 14 所示。

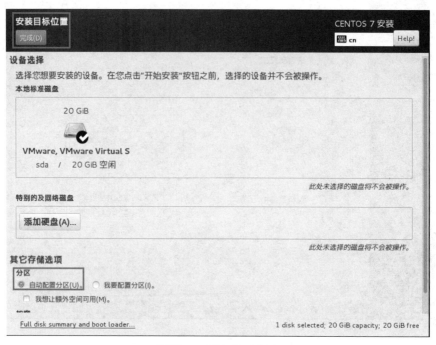

附图 14 硬盘分区设置

（11）以上安装选项设置好后即可进行最终的安装，附图 15 所示是桌面版安装，附图 16 所示是 mini 版安装。

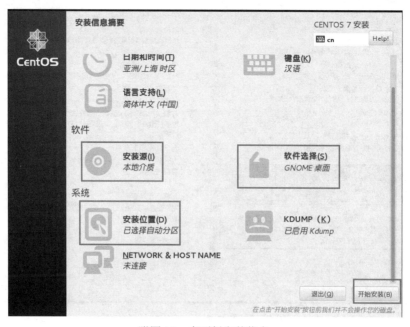

附图 15 桌面版安装信息

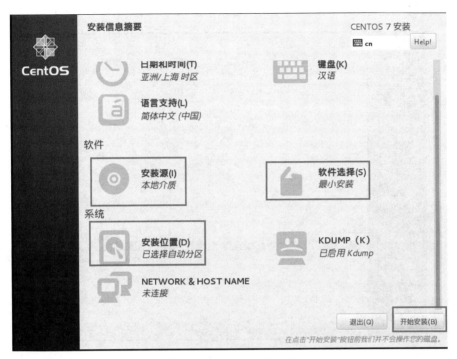

附图 16　mini 版安装信息

（12）在安装的过程中，需要设置 root 用户密码，密码设置好后单击两次"完成"按钮即可，如附图 17 所示。

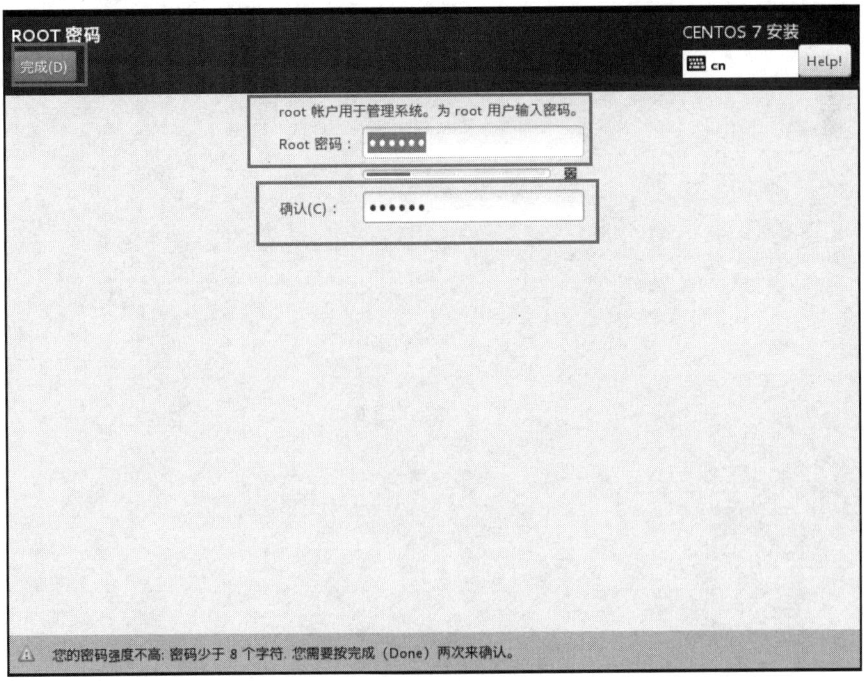

附图 17　root 用户密码设置

（13）安装完成后单击"重启"按钮即可进入 CentOS 操作系统，如附图 18 所示。

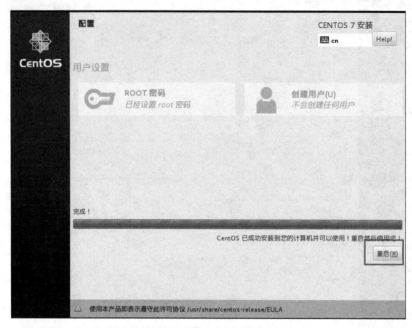

附图 18　CentOS 安装完成